# Introduction to Modern Climate Change, Second Edition

This is an invaluable textbook for any introductory survey course on the science and policy of climate change, for both non–science majors and introductory science students. The second edition has been thoroughly updated to reflect the most recent science from the latest IPCC reports, and many illustrations include new data. The new edition also reflects advances in the political debate over climate change. Unique among textbooks on climate change, this text combines an introduction to the science with an introduction to economic and policy issues, and it focuses closely on anthropogenic climate change. It contains the necessary quantitative depth for students to properly understand the science of climate change. It supports students in using algebra to understand simple equations and to solve end-of-chapter problems. Supplementary online resources include a complete set of PowerPoint figures for instructors, solutions to exercises, videos of the author's lectures, and additional computer exercises.

**Andrew Dessler** is a climate scientist who studies both the science and politics of climate change. His scientific research revolves around climate feedbacks, in particular how water vapor and clouds act to amplify warming from the carbon dioxide that human activities emit. During the last year of the Clinton administration, he served as a senior policy analyst in the White House Office of Science and Technology Policy. Based on his research and policy experience, he has authored two books on climate change: this textbook and *The Science and Politics of Global Climate Change: A Guide to the Debate* (co-written with Edward Parson; second edition published in 2010). This textbook won the 2014 American Meteorological Society Louis J. Battan Author's Award. In recognition of his work on outreach, in 2011 he was named a Google Science Communication Fellow. He is presently a professor of atmospheric sciences at Texas A&M University. His educational background includes a B.A. in physics from Rice University and a Ph.D. in chemistry from Harvard University. He also undertook postdoctoral work at NASA's Goddard Space Flight Center and spent nine years on the research faculty of the University of Maryland.

# Introduction to Modern Climate Change

**Second Edition**

ANDREW DESSLER

Texas A&M University

CAMBRIDGE UNIVERSITY PRESS

# CAMBRIDGE
## UNIVERSITY PRESS

University Printing House, Cambridge CB2 8BS, United Kingdom

One Liberty Plaza, 20th Floor, New York, NY 10006, USA

477 Williamstown Road, Port Melbourne, VIC 3207, Australia

314-321, 3rd Floor, Plot 3, Splendor Forum, Jasola District Centre, New Delhi - 110025, India

79 Anson Road, #06-04/06, Singapore 079906

Cambridge University Press is part of the University of Cambridge.

It furthers the University's mission by disseminating knowledge in the pursuit of
education, learning, and research at the highest international levels of excellence.

www.cambridge.org
Information on this title: www.cambridge.org/9781107480674

© Andrew Dessler 2016

First published 2016
11th printing 2020

Printed in the United Kingdom by TJ International Ltd. Padstow Cornwall

*A catalog record for this publication is available from the British Library.*

*Library of Congress Cataloging in Publication Data*
Dessler, Andrew Emory.
Introduction to modern climate change / Andrew Dessler, Texas A&M University. – [Second edition].
    pages   cm
Includes bibliographical references and index.
ISBN 978-1-107-09682-0
1. Climatic changes.   2. Climatic changes – Government policy.    I. Title.
    QC903.D46   2016
    551.6–dc23        2015014701

ISBN 978-1-107-09682-0 Hardback
ISBN 978-1-107-48067-4 Paperback

Additional resources for this publication at www.andrewdessler.com

For Michael and Alex

# Contents

# Preface

Future generations may well view climate change as the defining issue of our time. The worst-case scenarios of climate change are truly terrible, but even middle-of-the-road scenarios portend environmental change without precedent for human society. When future generations look back on our time in charge of the planet, they will either cheer our foresight in dealing with this issue or curse our lack of it.

Yet despite the stakes, the world has done basically nothing to address this risk. The reasons are obvious: The threat of climate change is really a threat to future generations, not the present one, so actions taken by our generation will mostly benefit them and not us. Moreover, such actions may be expensive – reducing emissions means rebuilding our energy infrastructure, and we have no idea how much that will cost. In such a situation, it is easiest to do nothing and wait for disaster to strike – which is why dams are frequently built after the flood, not before. Nevertheless, pushing this problem off onto future generations is a poor strategy. The impacts of climate change are global and mainly irreversible; by the time we have unambiguous evidence that the climate is changing and its impacts are serious, it will be too late to avoid these serious impacts. The only hope that future generations have to avoid serious climate change is us.

I fully believe that the cornerstone of good policy is an electorate that is educated on the issues, and this belief provided me the motivation for writing this book. The goal of this book is to cover the human-induced climate change problem from stem to stern, covering not just the physics of climate change but also the economic, policy, and moral dimensions of the problem. This sets it apart from most other climate change books, which typically do not have a tight focus on human-induced climate change or do not cover the nonscience aspects of the problem.

Such complete coverage of the climate change problem is essential. The science clearly underlies all discussion of the problem, and an understanding of the science is essential to an understanding of why so many people are so worried about it. Climate change, however, is no longer just a scientific problem. Virtually every government in the world now accepts the reality of climate change, and the debate has, to a great extent, moved on to policy questions, including the economic and ethical issues. Thus, one must also understand nonscience aspects of the problem to be truly informed on this issue.

The first seven chapters of the book focus on the science of climate change. Chapter 1 defines the problem and provides definitions of weather, climate, and climate change. It also addresses an issue that most textbooks do not have to address: why the reader should believe this book as opposed to Web sites and other sources that give a completely different view of the climate problem. Chapter 2 explains the evidence that the Earth is warming. The evidence is so overwhelming that there is little argument anymore over this point, and my goal is for readers to come away from the chapter understanding this.

Chapter 3 covers the basic physics of electromagnetic radiation necessary to understand the climate. I use familiar examples in this chapter, such as glowing metal in a blacksmith shop and the incandescent light bulb, to help the reader understand these important concepts. In Chapter 4, a simple energy-balance climate model is derived. It is shown how this simple model successfully explains the Earth's climate as well as the climates of Mercury, Venus, and Mars. Chapter 5 covers the carbon cycle, and feedbacks, radiative forcing, and climate sensitivity are all discussed in Chapter 6. Finally, Chapter 7 explains why scientists are so confident that humans are to blame for the recent warming that the Earth has experienced.

Chapter 8 begins an inexorable shift from physics to nonscience issues. It discusses emissions scenarios and the social factors that control them, as well as what these scenarios mean for our climate over the next century. Chapter 9 covers the impacts of these changes on humans and on the world in which we live. Chapter 10 covers exponential math. Exponential growth is a key factor in almost all fields of science, as well as in real life. In this chapter, I cover the math of exponential growth and explain the concept of exponential discounting. I also touch briefly on the social cost of carbon.

Starting with Chapter 11, the discussion is entirely on the policy aspects of the problem. Chapter 11 discusses the three classes of responses to climate change, namely adaptation, mitigation, and geoengineering, and their advantages, disadvantages, and trade-offs. The most contentious arguments over climate change policy are over mitigation, and Chapter 12 discusses in detail the two main policies advanced to reduce emissions: carbon taxes and cap-and-trade systems.

Chapter 13 provides a brief history of climate science and a history of the political debate over this issue, including discussions of the United Nations' Framework Convention on Climate Change and the Kyoto Protocol. Finally, Chapter 14 pulls the last three chapters together by discussing how to decide which of our options we should adopt, particularly given the pervasive uncertainty in the problem.

Overall, it should be possible to cover each chapter in three hours of lecture. This makes it feasible to cover the entire book in one fifteen-week semester. At Texas A&M, the material in this book is being used in a one-semester class for nonscience majors that satisfies the university's science distribution requirement. Thus, it is appropriate for undergraduates with any academic background and at any point in their college career.

Any serious understanding of climate change must be quantitative. Therefore, the book assumes a knowledge of simple algebra. No higher math is required. The book also assumes no prior knowledge of any field of science, just an open mind and a willingness to learn. To aid in the student's development of a numerate understanding of the climate, there are quantitative questions at the end of many of the chapters, and every chapter also has more open-ended, qualitative questions. In addition, there is a chapter summary at the end of each chapter that reviews and summarizes the most important takeaway messages from the chapter. A list of important terms is also provided at the end of each chapter. I've put additional readings, video recordings of my lectures, and computer exercises on my Web site, www.andrewdessler.com.

This is not an advocacy book. This is not to say that I do not have opinions. I do, and strong ones. I recognize, though, that shrill advocacy is frequently less effective than a dispassionate presentation of the facts. Thus, my strategy in this book is to simply explain the science and then lay out the possible solutions and trade-offs among them. I firmly believe that an unbiased assessment of the facts will bring the majority of people to see things the way I do: that climate change poses a serious risk and that we should therefore be heading off that risk by reducing our emissions of greenhouse gases.

Every year that our society does nothing to address climate change makes solving the problem both harder and more expensive. I am still optimistic, though, because problems often appear intractable at first. In the 1980s, as evidence mounted that industrial chemicals were depleting the ozone, it was not at all clear that we could avoid serious ozone depletion at a reasonable cost. The chemicals causing the ozone loss, namely chlorofluorocarbons, played an important role in our everyday life – in refrigeration, air conditioning, and many industrial processes – just like the main cause of climate change, fossil fuels, also plays an important role in our society. But the cleverness of humans prevailed. A substitute chemical was developed and it seamlessly and cheaply replaced the ozone-destroying halocarbons – at a cost so low that hardly anyone noticed when the substitution took place.

Solving the climate change problem will be harder than solving the ozone depletion problem – how much harder, no one knows. I am confident, though, that the ingenuity and creativeness of humans is such that we can solve this problem without damaging our standard of living. However, there is only one way to find out, and that is to try to do it.

# Acknowledgments

This book could not have been written without the incredible work of the climate science community. Ignored by many, demonized by some, I believe that future generations will look back and say, "They nailed it." I hope this book does justice to all of our hard work. The first edition of the book was written while I was on faculty development leave from Texas A&M University during Fall 2010. I thank the university for this support.

# 1     An introduction to the climate problem

We begin our trip through the climate problem by defining weather, climate, and climate change and by demonstrating how we use latitude and longitude to describe locations on the Earth. We also discuss something that few textbooks address: why you should believe this book.

## 1.1 What is climate?

The American Meteorological Society defines *climate* as

> The slowly varying aspects of the atmosphere–hydrosphere–land surface system. It is typically characterized in terms of suitable averages of the climate system over periods of a month or more, taking into consideration the variability in time of these averaged quantities.

Mark Twain, in contrast, famously summed it up a bit more concisely:

> Climate is what you expect; weather is what you get.

Put another way, *weather* refers to the actual state of the atmosphere at a particular time. Weather is what we mean when we say that, at 10:53 AM on November 15, 2014, the temperature in College Station, Texas, was 8 °C, the humidity was 66 percent, winds were out of the southeast at 8 knots, the barometric pressure was 30.23 inches, and there was no precipitation.

*Climate*, in contrast, is a statistical description of the weather over a period of time, usually a few decades. It would almost certainly include average temperature as well as a measure of how much the temperature varies about this average value, such as the record high and low temperatures. Figure 1.1 demonstrates one way to look at the climate: It shows the distribution of daily average temperatures in August near Fairbanks, Alaska, for two time periods, 1900–1929 and 1970–1999. During the 1900–1929 period, for example, the most likely daily average temperature was 10 °C, which occurred on approximately 16 percent of the days. Extremes occur less frequently; for example, the probability of temperatures above 16 °C or below 3 °C are small. The climate tells us only the range of probable conditions on a particular day; it contains no information about what the temperature was on any particular day.

In this book, I frequently use the Celsius scale, the standard temperature scale throughout the world (the Fahrenheit scale more familiar to U.S. readers is only used in the United States and a few other countries). For readers who may not be conversant

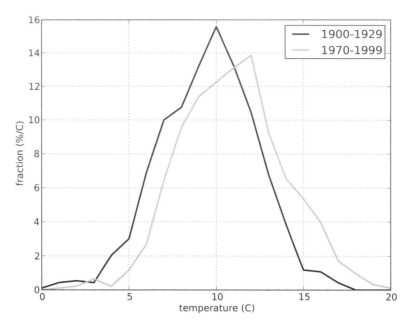

**Figure 1.1** Frequency of occurrence of daily average temperature in August at 64°N, 150°W, near Fairbanks, AK, for two time periods: 1900–1929 and 1970–1999 (data obtained from the twentieth-century reanalysis, version 2, www.esrl.noaa.gov/psd/data/gridded/data.20thC_ReanV2.html).

in Celsius, you can convert from Fahrenheit to Celsius using the equation $C = (F - 32) \times 5/9$; or from Celsius to Fahrenheit, $F = C \times 9/5 + 32$. It is also useful to remember that the freezing and boiling temperatures for water on the Celsius scale are 0°C and 100°C, respectively. On the Fahrenheit scale, these temperatures are 32°F and 212°F. Room temperature is about 22°C, which corresponds to 72°F.

Why do we care about weather and climate? Weather is important for making short-term decisions. For example, should you take an umbrella when you leave the house tomorrow? To answer this question, you do not care at all about the average precipitation for the month, but rather whether it is going to rain *tomorrow*. If you are going skiing this weekend, you care about whether new snow will fall before you arrive at the ski lodge and what the weather will be while you are there. You do not care how much snow the lodge gets on average.

Climate, however, is more important for long-term decisions. If you are looking to build a vacation home, you are interested in finding a place that frequently has pleasant weather – you are not particularly interested in the weather on any specific day. Plots like Figure 1.1 can help make these kinds of climate-related decisions; the plot tells us, for example, that a house in this location rarely needs air conditioning. If you are building a ski resort, you want to place it in a location that, on average, gets enough snow to produce acceptable ski conditions. You do not care if snow is going to fall on a particular weekend, or even what the total snowfall will be for a particular year.

An example of the importance of both the climate and the weather can be found in the planning for D-Day, the invasion of the European mainland by the Allies during World War II. The invasion required Allied troops to be transported onto the beaches of Normandy, along with enough equipment that they could establish and hold a beachhead. As part of this plan, Allied paratroopers were to be dropped into the French countryside the night before the beach landing in order to capture strategic towns and bridges near the landing zone, thus hindering an Axis counterattack.

There were important weather requirements for the invasion. The nighttime paratrooper drop demanded a cloudless night as well as a full moon so that the paratroopers would be able to land safely and on target, and then achieve their objectives – all before dawn. The sky had to remain clear during the next day so that air support could see targets on the ground. For tanks and other heavy equipment to be brought onshore called for firm, dry ground, so there could be no heavy rains just prior to the invasion. Furthermore, the winds could not be too strong because high winds generate big waves that create problems for both the paratroopers and the small landing craft that would ferry infantry to the beaches.

Given these and other weather requirements, analysts studied the climate of the candidate landing zones to find those beaches where the required weather conditions occurred most frequently. The beaches of Normandy were ultimately selected in part because of its favorable climate (tactical considerations obviously also played a key role).

Once the landing location had been selected, the exact date of the invasion had to be chosen. For this, it would not be the climate that mattered but rather the weather on a particular day. Operational factors such as the phase of the tide and the moon provided a window of three days for a possible invasion: June 5, 6, and 7, 1944. June 5 was initially chosen, but on June 4, as ships began to head out to sea, bad weather set in at Normandy, and General Dwight D. Eisenhower made the decision to delay the invasion. On the morning of June 5, chief meteorologist J. M. Stagg forecasted a break in the weather, and Eisenhower decided to proceed. Within hours, an armada of ships set sail for Normandy. That night, hundreds of aircraft carrying tens of thousands of paratroopers roared overhead to the Normandy landing zones.

The invasion began just after midnight on June 6, 1944, when British paratroopers seized a bridge over the Caen Canal. At dawn, 3,500 landing craft hit the beaches. Stagg's forecast was accurate and the weather was good, and despite ferocious casualties, the invasion succeeded in placing an Allied army on the European mainland. This was a pivotal battle of World War II, marking a turning point in the war. Viewed in this light, Stagg's forecast may have been one of the most important in history.

Temperature is the parameter most often associated with climate, and it is something that directly affects the well-being of the Earth's inhabitants. The statistic that most frequently gets discussed is average temperature, but temperature extremes also matter. For example, it is heat waves – prolonged periods of excessively hot weather – rather than normal high temperatures that kill people. In fact, heat-related mortality is the leading cause of weather-related death in the United States, killing many more people than cold temperatures do. And the numbers can be staggering: In August 2003, a severe heat wave in Europe lasting several weeks killed tens of thousands of people.

Precipitation rivals temperature in its importance to humans, because human life without fresh water is impossible. As a result, precipitation is almost always included in any definition of climate. Total annual precipitation is obviously an important part of the climate of a region. However, the distribution of this rainfall throughout the year also matters. Imagine, for example, two regions that get the same total amount of rainfall each year. One region gets the rain evenly distributed throughout the year, whereas the other region gets all of the rain in one month, followed by eleven rain-free months. The environment of these two regions would be completely different. Where the rain falls continuously throughout the year, we would expect a green, lush environment. Where there are long rain-free periods, in contrast, we expect something that looks more like a desert.

Other aspects of precipitation, such as its form (rain versus snow), are also important. In the U.S. Pacific Northwest, for example, snow that accumulates in the mountains during the winter melts during the following summer, thereby providing fresh water to the environment during the otherwise dry summers. If warming causes wintertime precipitation to fall as rain rather than snow, then it will run off immediately and not be available during the following summer. This can lead to water shortages during the summer.

As these examples show, climate includes many environmental parameters. What part of the climate matters will vary from person to person, depending on how each relies on the climate. The farmer, the ski resort owner, the resident of Seattle, and Dwight D. Eisenhower are all interested in different meteorological variables, and thus may care about different aspects of the climate. But make no mistake: We all rely on the stability of our climate. In particular, food production and freshwater availability, two of the most important things we rely on to survive, are greatly affected by the climate. I discuss this in greater depth when I explore climate impacts in Chapter 9.

A final difference between weather and climate is how easy they are to determine. Measuring the weather is pretty easy – just walk outside and look around.[1] If you need a higher level of accuracy, you can buy reasonably cheap instruments to measure the temperature, precipitation, or any other variable of interest. Climate, in contrast, is much harder to measure; it requires the gathering of decades of data so that we have sufficiently good, robust statistics, such as I plotted in Figure 1.1.

## 1.2 What is climate change?

The climate change that is most familiar is the seasonal cycle: the progression of seasons from summer to fall to winter to spring and back to summer, during which most non-tropical locations experience significant temperature variations. Precipitation can also vary by season. In fact, almost any climate variable can vary over the course of the year.

The concern in the climate change debate – and in this book – is with long-term climate change. The American Meteorological Society defines the term *climate change* as "any systematic change in the long-term statistics of climate elements (such as temperature, pressure, or winds) sustained over several decades or longer." In other words, we can compare the statistics of the weather for one period against those for another period, and if the statistics have changed, then we can say that the climate has changed.

---

[1] There are, of course, siting issues in measuring the weather. Depending on your location, the weather you measure when you walk outside may not be terribly representative of the weather of the larger areas.

Thus, we are interested in whether today's climate (defined over the past few decades) is different from the climate of a century ago, and we are worried that the climate at the end of the twenty-first century will be quite different from that of today. To illustrate this, Figure 1.1 shows the August temperature near Fairbanks, Alaska, for two periods, 1900–1929 and 1970–1999. Clearly, the temperature distributions in these two periods are different – the temperature distribution at the end of the twentieth century is about 2 °C warmer than at the beginning of the century. In other words, *the climate of this region has changed*. It should also be noted that there is no information on what caused the change – it may be due to global warming or any number of other physical processes. All we have identified here is a shift in the climate.

The shift in the temperature distribution is only ~2 °C, and it might be tempting to dismiss this as unimportant. However, as I discuss in Chapter 9, seemingly small changes in climate are associated with significant impacts on the environment. So you should not dismiss such a change lightly.

In Chapter 2, we will look more closely at data to determine if the climate is indeed changing. Before we get to that, however, there are two things I need to cover. First, in the next section, I discuss the coordinate system I will be using in this book.

## 1.3 A coordinate system for the Earth

I will be talking a lot in this book about the Earth, so it makes sense to define the terminology used to identify particular locations and regions on the Earth.

To begin, the *equator* is the line on the Earth's surface that is halfway between the North and South Poles, and it divides the Earth into a northern hemisphere and a southern hemisphere. The *latitude* of a particular location is the distance in the north-south direction between the location and the equator, measured in degrees (Figure 1.2). Latitudes for points in the northern hemisphere have the letter $N$ appended to them, with $S$ appended to points in the southern hemisphere. Thus, 30°N means a point on the Earth that is 30° north of the equator, whereas 30°S means the same distance south of the equator.

The *tropics* are conventionally defined as the region from 30°N to 30°S, and this region covers half the surface area of the planet. The *mid-latitudes* are usually defined as the region from 30° to 60° in both hemispheres, and these regions occupy roughly one-third of the surface area of the planet. The *polar regions* are typically defined to be 60° to the pole, and these regions occupy the remaining one-sixth of the surface area of the planet. The North and South Poles are located at 90°N and 90°S, respectively.

Latitude gives the north-south location of an object, but to uniquely identify a spot on the Earth, you need to know the east-west location as well. That is where *longitude* comes in (Figure 1.3). Longitude is the angle in the east or west direction, from the *prime meridian*, a line that runs from the North Pole to the South Pole through Greenwich, England, and is arbitrarily defined to be 0° longitude. Locations to the east of the prime meridian are in the eastern hemisphere and have the angle appended with the letter $E$, whereas locations to the west are in the western hemisphere and have the letter $W$ appended. In both directions, longitude increases to 180°, where east meets west at the international date line.

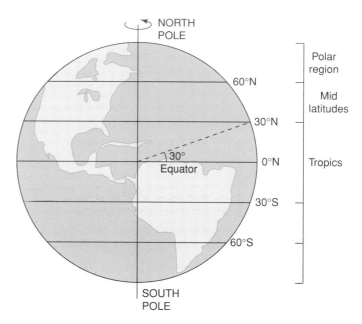

Figure 1.2    A schematic plot of latitude.

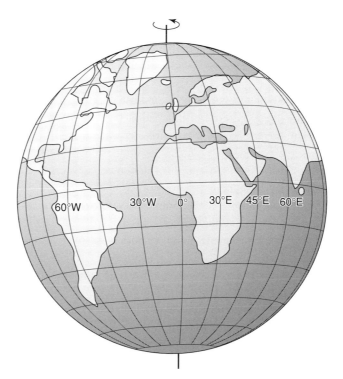

Figure 1.3    A schematic plot of longitude.

Together, latitude and longitude identify the location of every point on the planet Earth. For example, my office in the Department of Atmospheric Sciences of Texas A&M University is located at 30.6178°N, 96.3364°W. Knowing your location can literally be a matter of life and death – shipwrecks, wars, and other miscellaneous forms of death and disaster have occurred because people did not know where they were. Luckily for us, GPS (global positioning system) technology, which is probably built into your cell phone, can determine your latitude and longitude to within a few feet.

## 1.4 Why you should believe this textbook

I now have to address an issue that generally does not come up in college textbooks: why you should believe it. Students in most classes accept without question that the textbook is correct. After all, the author is probably an authority on the subject, the publisher has almost certainly reviewed the material for accuracy, and the instructor of the class, someone with knowledge of the field, selected that textbook. Given those facts, it seems reasonable to simply assume that the information in the textbook is basically correct.

But climate change is not like every other subject. If you do a quick Internet search, you will be able to find a Web page that disputes almost every claim made in this textbook. Your friends and family may not believe that climate change is a serious problem, or they may even believe it is a hoax. You may agree with them. This book will challenge many of these so-called skeptical viewpoints, and you may face the dilemma of whom to believe.

This situation brings up an important and interesting question: How do you determine whether or not to believe a scientific claim? If you happen to know a lot about an issue, you can reach your own conclusions on the issue. However, no one can be an expert on every subject; for the majority of issues on which you are not an expert, you need a shortcut.

One type of shortcut is to rely on your firsthand experience about how the world works. Claims that fit with your own experience are easier to accept than those that run counter to it. People do this sort of evaluation all the time, usually unconsciously. Consider, for example, a claim that the Earth's climate is not changing. In your lifetime, climate has changed very little, so this seems like a plausible claim. However, a geologist who knows that dramatic climate shifts are responsible for the wide variety of rock and fossil deposits found on Earth might regard the idea of a stable climate as ludicrous, but in turn might be less likely to accept a human origin for climate change. The problem with relying on firsthand experience about the climate is that our present situation is unique – people have never changed the composition of the global atmosphere as much or as fast as is currently occurring. Thus, whatever the response will be, it will likely be outside the realm of our and the Earth's experiences.

Another type of shortcut is to rely on your values: You can accept the claims that fit with your overall worldview while rejecting the claims that do not. For example, consider the scientific claim that secondhand smoke has negative health consequences. If you are a believer in unfettered freedom, you might choose to simply reject this claim out of hand because it implies that governments should regulate smoking in public places to protect public health. Those who are more suspicious of the integrity of big business are going to be more skeptical of the efficacy of vaccines because they believe that corporations are willing to put profits ahead of safety.

Yet another shortcut is to rely on an *opinion leader*. Opinion leaders are people who you trust because they appear to be authoritative or because you agree with them on other issues. They might include a family member or influential friend, a media figure such as talk show hosts Rush Limbaugh and Jon Stewart, or an influential politician such as Barack Obama or George W. Bush. In the absence of a strong opinion of your own, you can simply adopt the view of your opinion leaders. The problem with this approach is that there is no guarantee that the opinion leaders have a firm grasp of the science.

The most widely accepted approach is to rely on the opinions of experts. When the relevant experts on some subject have high confidence that a scientific claim is true, that is the best indication we have that the claim actually is true. This is not just my view; I am willing to bet it is something you believe in, too. If a friend tells you that he thinks he may be sick, what would you recommend? Your recommendation is likely to be that he should go see a doctor – and not just any doctor, but one who is an expert in that particular ailment.

This is also the view of the U.S. legal system. Many court cases involve questions of science (e.g., what was the cause of death, does a particular chemical cause cancer, does a DNA sample match the defendant). To settle those cases, the court will frequently turn to expert witnesses. These expert witnesses are, as their name suggests, experts on the matter that they are testifying about, and they provide relevant expertise to the court to help evaluate the important scientific questions that a case may revolve around.

To be an expert witness, one must demonstrate expertise in a particular subject. I have served as an expert witness on climate change in lawsuits over the permitting of coal-fired power plants, and the court qualifies me as an "expert" by using my research in climate change as well as the textbooks I have authored as evidence.

It should be emphasized that one must demonstrate specific, recent expertise in the exact area under consideration to be an expert witness. Showing expertise in general technical matters or in a related field is not sufficient. For example, one might consider a scientist with a Ph.D. in another field (e.g., solid-state physics) to have a credible opinion about the science of climate change. This is not so, and a person with a Ph.D. but without specialized knowledge of the climate would not qualify as an expert on matters of climate. That also goes for weather forecasters – climate and weather are different, and being an expert in weather does not qualify someone to be an expert witness on climate. And the requirement for the expertise to be recent rules out those who were experts, say, a decade ago but who have not kept up with the latest discoveries in the field.

There are many more examples that demonstrate that, as individuals and as a society, we rely on experts when evaluating complex technical issues. That is probably a good thing, too, because on a planet with 7 billion people, you can always find someone who will contest any claim, no matter how well established it is. For example, it would be relatively easy to find someone somewhere who would dispute the claim that cigarettes cause health problems. So if everyone's opinion counted equally, then it would be impossible to ever settle any dispute over a scientific claim – even one as simple as whether the Earth goes around the Sun.

But we also know that even the most trusted expert can be wrong, so the opinion of a single expert should be taken with caution. One way to gain confidence in a particular expert opinion is to ask several experts instead of just one. We frequently do this for important medical decisions by getting a second opinion. For high-stakes medical decisions, you would ideally solicit the opinions of many experts. If all of these experts were to agree, then you would have justifiably high confidence that the recommendations are the best advice that modern medicine can provide.

Climate change is really no different. It is obvious that the relevant experts are the community of climate scientists. And rather than listen to any single individual, we would do best by asking a large, representative sample of the world's climate scientists what they think – and if the vast majority agree on a particular point, then we can have high confidence that this is best estimate science can provide.

This is, in fact, what has already been done. In 1988, as nations began to acknowledge the seriousness of the climate problem, the Intergovernmental Panel on Climate Change (IPCC) was formed. The IPCC assembles large writing teams of scientific experts and has them write, as a group, reports detailing what they know about climate change and how confidently they know it. The reliance on large writing groups reduces the possibility that the erroneous opinions of an individual or a small group make it into the report, much like getting multiple opinions in medicine reduces the chance of a bad diagnosis.

To further minimize the possibility that the group of scientists writing the report are biased in some direction, the scientists making up the writing teams are not assembled by a single person or organization; they are nominated by the world's governments. Thus, the only way the IPCC's writing groups would be biased in some direction is if all of the world's governments nominated appropriately biased individuals. This seems unlikely, particularly since addressing climate change brings a raft of short-term problems to most governments. Many governments would therefore be happy if climate change disappeared completely as a political issue and therefore have no incentive to nominate scientists biased to the view climate change as a serious problem.

Drafts of the IPCC's reports are reviewed prior to release by other expert scientists, and they undergo a public review and a separate review by the world's governments. In the end, the IPCC's reports are widely regarded as the most authoritative statements of scientific knowledge about climate change, and as such they carry enormous weight in both the scientific and the policy communities. In 2007, the IPCC shared the Nobel Peace Prize in recognition of its work on the climate.

---

### An aside: The Summary for Policymakers

If you have ever tried to read an IPCC report, you know that the 1,000 plus page reports can be baffling for non-experts. That is why every report also has a *Summary for Policymakers*, a more readable summary of the full report that runs a few dozen pages. Referred to as the "SPMs," they summarize in more general language the most important conclusions in the main report.

The SPMs also serve another unique function. During a final meeting after the main report is written, representatives from each of the world's governments review a draft SPM written by scientists and vote on every sentence. Only if there is unanimous agreement from all of the world's governments is a sentence included in the SPM. During this process, sentences are frequently rewritten to make them acceptable to the world's governments. If there is nearly unanimous agreement on a sentence, with just one or two countries dissenting, than the sentence can be included in the SPM with a footnote recording the dissent.

The purpose of this exercise is to produce a common set of scientific facts to serve as the basis of future negotiations on policy. By having unanimous agreement on every sentence, no country can later say during policy negotiations that they don't agree with a particular scientific fact – they have already agreed to everything in the SPM.

This means, though, that every country is also trying to mold the SPM to best suit their negotiating position. During the meeting for the SPM for the IPCC's 1995 report, for example, Saudi Arabia and Kuwait argued strenuously to weaken the statements about humans causing climate change. When the rest of the world disagreed, it was then proposed that a footnote would be added to the report noting the disagreement – but the footnote was removed at Saudi Arabia and Kuwait's request because it would have been embarrassing for those two major oil producers to be the only countries in the world to not accept the scientific evidence of human impacts on climate.

In the end, the SPM represents a good summary of our scientific understanding of the climate but one that has an unavoidable hint of political influence in it. To the extent that political wrangling affects the SPM, it is almost always to water down the conclusions – reduce our confidence in scientific statements, lessen the impacts, etc. But despite these flaws, the SPMs should be given considerable deference in policy debates over climate.

---

Despite the careful process that produces the IPCC reports, it remains controversial in the public debate. To understand why, consider the curious case of cigarettes. Scientists have known since the 1950s that smoking cigarettes is terrible for your health. And in 1964, a landmark report by the U.S. surgeon general laid out the evidence in great detail for the general public. The tobacco companies at that point had two choices: they could accept that their product killed their customers, an admission that would certainly reduce their profits, or they could fight back by attacking the science.

Perhaps unsurprisingly, they chose to fight the science. In a memo released just after the surgeon general's report, a tobacco executive plotted out the response: "We must in the near future provide some answers which will give smokers a psychological crutch and a self rationale to continue smoking. These answers must also point out the weaknesses in the [Surgeon General's] Report."[2]

---

[2] From the Legacy Tobacco Documents Library, legacy.library.ucsf.edu/tid/ctv74e00/pdf;jsessionid=F68A45A37FAF5E3AD23A5E84A7EEE463.tobacco03

Good to their word, the tobacco companies proceeded to wage a successful multi-decade campaign to cast doubt on the science. It was only in the 1990s, four decades after the science was actually settled, that the phony public debate over the health impacts of smoking finally faded away.

Today, the tobacco debates are the archetype of the dishonest manipulation of science in pursuit of a particular policy goal.[3] And the dishonesty paid handsomely: it effectively delayed by decades public awareness of the strength of the scientific consensus of the dangers of cigarettes, thereby keeping people smoking and delaying government action to restrict its sales. Given how successful the fight waged against tobacco science was, it should come as no surprise that those opposed to political action to address climate change have adopted a similar approach.

Echoing the tobacco debate, one argument frequently made during the climate debate is that there is, in fact, wide disagreement on the science of climate change among climate scientists. As evidence, they will point to Internet petitions and various lists of scientists that dispute the mainstream view. However, a close evaluation of the dissident scientists on these lists and petitions reveals that in almost all cases they should not be considered experts on climate. Although many of the individuals on the lists have technical degrees, and some even have doctorates, their specific training does not include climate change. They would never qualify as an expert witness in a lawsuit on climate change; they do not have the background to teach a college-level course on the material; and we would never trust the diagnosis provided by a doctor with equivalent expertise.

The only reason that advocates put such transparently unqualified people forward as experts is that legitimate experts with the desired opinions are not available. Thus, the lack of credentials of those on the petitions and lists actually underscores the strong agreement among the relevant experts on the science of climate change.

A second claim we may hear is that climate scientists are manufacturing a crisis to benefit themselves. If climate is a crisis, so the argument goes, then more research funding will flow into the field, the prestige of climate scientists will increase, and scientists will be able to implement their preferred social policies.

There is, in fact, no evidence to support this argument. Rather, this argument relies on the listener simply accepting the obviousness of the claim that an entire scientific field would be willing to engage in scientific misconduct for research funding. What is often lost in this discussion is that all scientific fields have this same incentive. Biomedical fields could invent a new disease or a cure for an existing one to increase funding; physicists studying solid-state physics could invent a discovery that could lead to much faster computers; and space physicists could invent evidence of life elsewhere in the solar system. All of these "discoveries" would increase funding and interest in the particular field of interest.

It turns out that there are strong barriers to such widespread fraud by an entire scientific community. First and foremost, there is a coordination problem: How do you get everyone to go along with the fraud? The answer is that you can not. A scientific field such as climate science is a large, diverse, and intensely competitive endeavor, and the desire to outthink one's peers and show that one is smarter than

---

[3] The movie "Thank You for Smoking" (www.imdb.com/title/tt0427944/) is a great parody of the debate.

they are is much greater than the incentive to cooperate in this type of fraud. The reason for this is that success in science is achieved by impressing one's colleagues. One of the best ways to do this is to overturn conventional wisdom, either by showing that previous scientific results are wrong or by suggesting a new theory that fits the data better. Because this is so beneficial to the reputation of the individual scientist, it provides a strong incentive against participating in any conspiracy.

And the incentives in science do not support such a conspiracy. Money from grants does not generally go into the pockets of the researchers. Most scientists are employees of federal or state governments or private universities, and the amount of money they can pay themselves off grants is extremely limited. Instead, the majority of grant money goes to buy equipment or pay for graduate students. Thus, research funding provides a very weak incentive to cheat.

Finally, the entire underlying premise of the "climate science is corrupt" narrative is questionable. The premise is that, by suggesting that human effects on the climate are well understood, the field gets more research funding than it would otherwise. However, history suggests that whenever a field reaches a conclusion on a problem, funding for further research on that problem goes down. For example, after ozone depletion was confidently attributed to ozone-destroying chemicals known as chlorofluorocarbons in the mid-1990s, the funding for subsequent research rapidly dropped. By saying that they understand the climate system reasonably well, members of the climate science community are *not* helping their funding. They would do better if they claimed that there was no consensus on why the climate is warming. In that case, it would almost certainly be a high priority for most policymakers to fund research into determining the cause of the warming.

But while there has never been widespread fraud by an entire scientific community, there have been cases in which advocates opposed to political action have falsely tried to cast doubt on the science. One such example was discussed earlier: leaked documents from tobacco companies clearly show the intent of these companies was to generate doubt in the public's mind about the health effects of cigarette smoking.

Because of this, we should be leery of the argument that "the experts can't be trusted." It goes against both common sense and our experience in the real world, and it should only be accepted if extraordinary evidence is provided. In the climate change debate, such evidence is clearly lacking.

## An aside: But I still don't trust the IPCC

If you talk to people about climate change, you will find some people who, no matter what arguments you make, will simply not accept the IPCC reports. To those people, you should point out the many other reports written by authoritative organizations, such as the U.S. National Academy of Sciences and the U.K. Royal Society.[4] Or you can look at the statements put out by the scientific societies that climate experts belong to. In October 2009, for example, a collection of U.S. scientific organizations sent a letter to the U.S. Senate stating that climate change is a serious problem facing

---

[4] See the additional reading for this chapter at www.andrewdessler.com/chapter1 for links to these reports and other examples.

the entire human race and that emissions of greenhouse gases have to be dramatically reduced for us to avoid the most severe impacts. Signatories of this letter include the American Association for the Advancement of Science, the American Chemical Society, the American Geophysical Union, the American Institute of Biological Sciences, the American Meteorological Society, the American Society of Agronomy, the American Society of Plant Biologists, the American Statistical Association, the Association of Ecosystem Research Centers, the Botanical Society of America, the Crop Science Society of America, the Ecological Society of America, the Natural Science Collections Alliance, the Organization of Biological Field Stations, the Society for Industrial and Applied Mathematics, the Society of Systematic Biologists, the Soil Science Society of America, and the University Corporation for Atmospheric Research. Comparable non-U.S. scientific organizations in other countries have also endorsed the mainstream view of the science of climate change.

The science in this book follows the reports of the IPCC and the other relevant national and international scientific organizations. That, in a nutshell, is why you should believe this book. The alternative views on climate change you might see or hear, such as those from skeptical friends or the Internet, do not come from a process as credible as the IPCC's and therefore should not have the same standing. It should be emphasized that this does not mean the IPCC's reports are correct – any scientific claim is at risk of being overturned by future research. Nevertheless, the IPCC's reports do accurately represent what the relevant experts think about the science, which is the best guide there is for non-experts.

## 1.5 Chapter summary

- *Weather* refers to the exact state of the atmosphere at a point in time; *climate* refers to the statistics of the atmosphere over a period of time, usually several decades in length or longer.
- Climate change refers to a change in the statistics of the atmosphere over decades. Such statistics include not just the averages but also the measures of the extremes – how much the atmosphere can depart from the average.
- Temperatures expressed in this book are in degrees Celsius; conversion from Fahrenheit can be done with this equation: $C = (F - 32) \times 5/9$.
- Any position on the surface of the Earth can be described by a latitude and longitude. Latitude is a measure of the position in the North-South direction, while longitude is a measure of the position in the East-West direction. The tropics cover the region from 30°N to 30°S; mid-latitudes cover the region from 30° to 60° latitude; and the polar regions cover from 60° to 90° latitude.
- In our society, we frequently rely on experts for advice on highly specialized or technical fields. For climate change, the IPCC reports represent the opinion of the world's experts, and the science described in this book reflects the IPCC's scientific views.
- Those opposed to policy action on climate change frequently make doubt of the science of climate change a central part of their argument. In so doing, they are

following the strategy employed by the tobacco companies to keep public debate about the health effects of smoking alive for decades after the issue was settled by scientists.

## Additional reading

The IPCC's reports are available online from www.ipcc.ch.

A. E. Dessler and E. A. Parson, *The Science and Politics of Global Climate Change: A Guide to the Debate*, 2nd ed. (Cambridge: Cambridge University Press, 2010). Chapter 2 discusses how the scientific and policy processes work and how assessments (like the IPCC) help bridge the gap between them.

N. Oreskes and E. M. Conway, *Merchants of Doubt: How a Handful of Scientists Obscured the Truth on Issues from Tobacco Smoke to Global Warming* (London: Bloomsbury Press, 2010). This is an important book about how deception is used to mislead the public on matters ranging from the risks of smoking to ozone depletion to the reality of global warming.

Union of Concerned Scientists, *Smoke, Mirrors, and Hot Air: How ExxonMobil Uses Big Tobacco's Tactics to Manufacture Uncertainty on Climate Science* (Cambridge, MA: Union of Concerned Scientists, January 2007). This is a description of how the tactics employed by the tobacco companies to cast doubt on the science of the health impacts of smoking are now being used by oil companies to cast doubt on the science of climate change (available online at www.ucsusa.org/assets/documents/global_warming/exxon_report.pdf).

Most scientific societies have statements affirming the science of climate change presented in this book. This includes statements from the American Geophysical Union and the American Meteorological Society.

See www.andrewdessler.com/chapter1 for links to the above material and additional resources for this chapter.

## Terms

Climate
Climate change
Equator
Latitude
Longitude
Mid-latitudes
Opinion leader
Polar region
Prime meridian
Summary for Policymakers
Tropics
Weather

# Problems

1. Determine the latitude and longitude of the White House, the Kremlin, the Pyramids of Giza, and the point on the opposite side of the Earth to where you were born. Use an online tool (e.g., Google Earth) or an atlas (which you can find in any library).

2. a) Convert the following temperatures from degrees Fahrenheit to degrees Celsius: 300, 212, 70, 50, 32, and 0°F
   b) Convert the following temperatures from degrees Celsius to degrees Fahrenheit: 150, 100, 70, 50, 0, and −10°C

3. a) The temperature increases by 1°C. How much does it increase in degrees Fahrenheit?
   b) The temperature increases by 1°F. How much does it increase in degrees Celsius?
   c) This is true: I told a reporter that the Earth has warmed by 0.8°C over the last century. When it appeared in print, the sentence said: Dessler said that the Earth has warmed by 33°F over the last century. Where did the reporter go wrong?

4. What temperature has a numerical value that is the same in degrees Celsius as it is in degrees Fahrenheit?

5. Find a two-digit temperature in degrees Fahrenheit for which, if you reverse the digits, you get that same temperature in degrees Celsius (e.g., find a temperature, such as 32°F, for which the Celsius equivalent would be 23°C; this example, of course, does not work).

6. Why do you believe that smoking causes cancer? (If you do not believe this, then why do you believe that smoking does not cause cancer?) What would be required to get you to adopt the opposing view?

7. Find two friends who have strong but opposing views of climate change.
   a) Ask both of them why they believe what they do and what would be required for them to adopt the opposing view. It is important to understand where their views come from; if they argue, say, that glaciers are retreating or not, find out where they get their facts.
   b) Which of these positions appears more credible? Why?
   c) Can you use their views on climate change to predict their views on other issues (abortion, gun control) and their political affiliation?

8. Practice reading a graph. These questions all refer to Figure 1.1.
   a) What fraction of days have an average temperature of 15°C during the 1900–1929 and the 1970–1999 periods?
   b) For the 1900–1929 period, what is the warmest temperature that occurred? What about the 1979–1999 period?
   c) What temperature(s) have an equal probability of occurring in the two periods?
   d) For the period 1970–1999, estimate the fraction of days that have a temperature 15°C *or greater*.
   e) Same as d, but for the period 1900–1929.
   f) Compare the answers to d and e. What does this tell you about the changes in extreme heat under even modest warming?

9. Give examples of situations when weather affected your or your family's life. Then do the same for climate.

10. Those opposed to the IPCC's scientific conclusions have set up their own summary of the science of climate change, which they call the NIPCC. Do some online research and then compare and contrast the credibility of the two reports.

11. In the climate debate, few institutions are attacked as frequently as the IPCC. Using web searches, identify some arguments made by those arguing that the IPCC cannot be trusted.

# 2

# Is the climate changing?

In this chapter, I address the question of whether the Earth's climate is currently changing and how it has changed in the past. You will see overwhelming evidence that the climate is indeed changing and that it has changed significantly over the Earth's entire history. I will not discuss the causes of climate change here, though – we will do that in Chapter 7.

In Chapter 1, climate change was defined as a change in the statistics of the weather over a period of several decades. In this chapter, the statistic I will primarily focus on is global average temperature for two reasons. First, the most direct impact from the addition of greenhouse gases to the atmosphere is an increase in temperature. Changes in other variables, such as precipitation or sea level, are a response to the temperature change. Second, we have the best data for temperature. The technology for measuring it is centuries old, and people have been measuring and recording the temperature with reasonable global coverage since the middle of the nineteenth century. In addition to direct temperature measurements, there are other techniques, such as studying the chemical composition of ice and rocks, that allow us to indirectly infer the temperature of the Earth over nearly its entire 4.5-billion-year history.

Rather than analyze temperature directly, however, we will instead analyze *temperature anomalies*, defined as the difference between the temperature and a reference temperature; the reference temperature is usually the average temperature over a previous multi-decadal period. For example, the monthly average temperature for Australia in August 2009 was 19.4°C, while the August 1979–2009 average for that location was 15.4°C. Thus, that region's temperature anomaly in August 2009 was +4.0°C, relative to the 1979–2009 baseline. Had we picked a different reference period, the anomaly would change. As an example, Figure 2.1 shows the global pattern of anomalies for June 2009.

Why use anomalies rather than absolute temperature? The main reason is that absolute temperature can vary sharply over short distances, such as between a city and a nearby rural area, or between two nearby sites at different altitudes. You may have noticed this, for example, if your car displays outside temperature on its dashboard. As you drive a few miles, you might see the temperature change by a few degrees, particularly if you are driving into and out of a city.

Anomalies, however, are constant over much longer distances: If it is a degree warmer than average in a city, then it will be a degree warmer than average a few kilometers away from the city, even if the absolute temperatures differ by a few degrees. Figure 2.1 shows this – regions of warm and cold anomalies tend to be hundreds, sometimes even thousands, of kilometers across. This means that calculations of global average temperature anomalies require only about a hundred or so temperature stations spread across the globe. Measuring the absolute

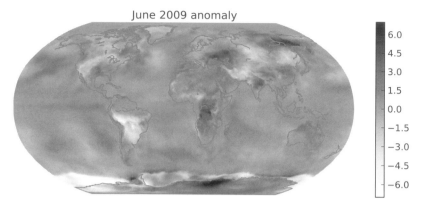

June 2009 anomaly

**Figure 2.1** The monthly surface temperature anomaly in June 2009 in degrees Celsius. The reference temperature for the anomaly is the average of the June temperatures from 1979 to 2009 (data obtained from the MERRA reanalysis). (See Color Plate 2.1.)

temperature of the planet would require many more stations – more stations, in fact, than exist.

Another advantage of using anomalies is that you can measure changes in a quantity even if you cannot measure the absolute value of the quantity. Imagine, for example, that you want to determine if a child is growing. The most obvious way to do this is to measure the child's height every few months. But, if you could not do that, an alternative would be to measure his height relative to, say, a mark on the wall. If the top of his head is 1 inch below the mark one year, even with it the next year, and 1 inch above the following year, then you can be confident that the child is growing 1 inch per year – even if you never know the child's absolute height. This is the situation for the Earth's temperature. We cannot measure the Earth's absolute temperature with high accuracy (because measuring it would require an extremely dense network of temperature measurements, as discussed earlier), but we can measure the temperature anomaly with high accuracy – high enough to see clear warming.

I will also generally focus in our discussion on global average quantities. The reason is that the climate of a region can vary significantly just due to weather variability – i.e., particular regions can experience climate extremes (e.g., a heat wave) that are completely independent of climate change. However, these local variations are usually balanced by an opposite extreme elsewhere: if one region is undergoing a heat wave, there is likely another region that is undergoing a cold wave (as seen in Figure 2.1). By averaging over the globe, we rid ourselves of most of this weather variability and more clearly isolate the smaller climate change signal.

## 2.1 Recent climate change

### 2.1.1 Surface thermometer record

People have been measuring the local air temperature at locations all over the globe for centuries. They were originally made manually using liquid-in-glass thermometers,

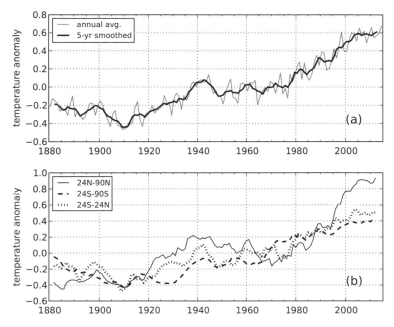

**Figure 2.2**    (a) Global annual average temperature anomaly; the gray line is the annual average, and the black line is a smoothed time series. (b) Smoothed temperature anomaly time series for three different regions of the planet: the northern hemisphere (24°N-90°N), the tropics (24°N-24°S), and the southern hemisphere (24°S-90°S). In both plots, the reference temperature used in calculating the anomalies is the 1951–1980 average. Data are from the NASA GISS Surface Temperature Analysis [Hansen et al., 2010], downloaded from data.giss .nasa.gov/gistemp/.

but in recent decades these have been replaced by automated electronic thermometers. By combining these measurements, scientists can estimate the global average surface temperature anomaly of the Earth over the past 150 years, and that time series is plotted in Figure 2.2a.

The data clearly show that the Earth is warming. From 1880 to 2012, the average surface temperature of the Earth rose by 0.85°C. The warming has not been uniform but occurred primarily in two distinct periods, from 1910 to 1945 and from 1976 to 2002. Superimposed on the slow warming trend are many bumps and wiggles that are unrelated to climate change. Despite this short-term variability, the recent warming is basically continuous, with every decade since the mid-twentieth century warmer than previous decades. The three warmest years in the record were 2005, 2010, and 2014.

It is also worth noting that the year-to-year variations in the global-average temperature are quite small – just a few tenths of a degree. This is much smaller than the local temperature variations where you live. Thus, warming of a few degrees Celsius, which are predicted for the twenty-first century, would run off the scale in Figure 2.2. While this does not tell us the effect of such warming, it should certainly compel our attention.

Figure 2.3 shows how the warming of the twentieth century was distributed across the planet. Clearly, the warming is occurring just about everywhere – thus justifying the "global" part of "global warming." However, the warming has not been entirely

Observed change in surface temperature 1901–2012

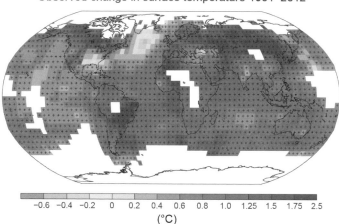

−0.6  −0.4  −0.2   0   0.2  0.4  0.6  0.8  1.0  1.25  1.5  1.75  2.5
(°C)

**Figure 2.3**   The distribution of warming (in °C) between 1901 and 2012. Regions where data are too sparse to produce an estimate are white. Adapted from Figure SPM.1 of IPCC [2013]. (See Color Plate 2.3.)

uniform. Probably the most obvious difference is that land areas warmed more than the ocean. Figure 2.2b shows temperatures for the northern hemisphere, tropics, and southern hemisphere, and it shows that the northern hemisphere warmed more than the tropics or the southern hemisphere and that the tropics and southern hemisphere warmed about equally.

In science, no single data set is ever considered definitive, and that is particularly true of the surface thermometer record. This network of thermometers was not designed for climate monitoring, and, over the years, the network has undergone many changes. Changes in the types of thermometer used, station location and environment, observing practices, and other sundry alterations all have the capacity to introduce spurious trends in the data.

For example, imagine you have a thermometer that is in a rural location in the late nineteenth century. Over time, a nearby city expands so that by the 1980s, the thermometer is completely surrounded by the city. Because cities tend to be warmer than nearby rural locations, this would introduce a warming trend in the data not caused by a warming climate.

Scientists know about these problems, and, to the extent possible, they adjust the data to take them into account. For example, the impact of a city growing up around a thermometer can be assessed by comparing the measurements from that thermometer to nearby thermometers that have remained rural for the entire period. The temperature record in Figure 2.2 includes adjustments to account for as many of these issues as possible.

Nevertheless, uncertainty in the data remains, as a result of both uncertainties in the adjustments and uncertainties that cannot be adjusted for. To account for this, scientists put *error bars* on the trend. An error bar is the scientists' estimate of the potential error in their estimate. The name "error bar" comes from the fact that, on a plot, error bars are frequently indicated as bars extending from the data. For the

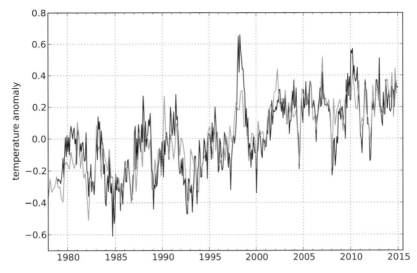

**Figure 2.4**   Satellite measurements of the global monthly average temperature anomaly (black line). The gray line is temperature from the surface thermometer record. Anomalies in this plot are relative to the 1981–2010 period (data obtained from the University of Alabama, Huntsville; downloaded from www.nsstc.uah.edu/data/msu/t2lt/uahncdc_lt_5.6.txt).

warming from 1880 to 2012 of 0.85°C, the error bar is 0.20°C, meaning that the warming is very likely between 0.65°C and 1.05°C.

Given all of the possible problems in these data, it would be foolish to rely entirely on this one source to determine if and how much the Earth was warming. Scientists therefore turn to other data sets to verify this result. In the rest of Section 2.1, I describe the other data sets used to build confidence in the surface thermometer data set.

## 2.1.2 Satellite measurements of temperature

It is possible to measure global average temperature from orbit, and the United States has been flying instruments on satellites to make that measurement since 1978. Figure 2.4 shows the time series of satellite measurements of the global monthly average temperature anomaly. These data show a general warming trend over this period of approximately 0.14°C per decade (1.4°C per century).

As with all data sets, though, this one has its own set of problems and uncertainties. First, satellites actually measure the average temperature of the lowest 8 km of the atmosphere, from the surface to about the altitude where airliners fly. Thus, it is not actually a measurement of the surface temperature, although the temperature of this layer of the atmosphere should track the surface temperature.

Another issue with these data is orbital drift of the satellites carrying the instruments. Imagine that a satellite flies over a location at 2 PM each day and makes a measurement of that location's temperature. Over time, the satellite's orbit drifts so that it flies over that location later and later each day. After a few years, the satellite is flying over that location at 3 PM. Because temperatures rise throughout the day, it is generally warmer at that location at 3 PM than it is at 2 PM. Thus, the drift in the satellite's orbit would by itself introduce a warming trend, even if the climate were not actually changing. This artifact must also be identified and adjusted for.

Other issues include calibration of the satellite instruments, which were never designed to make long-term measurements, and the shortness of the satellite record (just a few decades long), both of which also introduce uncertainty into the observed warming. As with the surface thermometer record, these issues are known and adjusted for, to the extent possible.

One way to gain confidence in the satellite and surface thermometer records is to compare them; this is done in Figure 2.4 (the same surface thermometer data, but annually averaged, was plotted in Figure 2.2). The excellent agreement between these two independent temperature measurements provides strong confirmation of the reality of the warming of the climate seen in both data sets.

It is also worth noting that superposed on the long-term warming trend in Figure 2.4 are lots of shorter-term ups and downs. These are not random but can be assigned to various physical causes, mainly El Niño-La Niña cycles and volcanic eruptions. During El Niño events, the Earth warms several tenths of a degree Celsius. El Niño's opposite is La Niña, and during those events the Earth cools several tenths of a degree. These El Niño-La Niña events cause temporary fluctuations in temperature lasting a few years but no long-term changes in the climate. Volcanic gases emitted during eruptions cool the climate by blocking sunlight – after a few years, the effluents are removed from the atmosphere and the climate returns to normal. These processes will be discussed in more detail later in the book.

## An aside: Has global warming stopped?

> "The planet has largely stopped warming over the past 15 years, data shows."
> – FoxNews.com, September 27, 2013[1]

One claim that has arisen recently in the public debate over climate change is that the Earth was warming, but it stopped warming five, ten, or fifteen years ago. The implication of this argument is that climate change is therefore nothing to worry about.

So has climate change stopped? Figure 2.4 showed that superimposed on the slow warming trend are lots of bumps and wiggles associated with random variability from things like weather (a particularly cold winter or a particularly warm summer) or volcanic eruptions (which cool the planet) or natural cycles like (El Niño-La Niña cycles).

These sources of short-term variability do not have anything directly to do with climate change and do not cause any long-term changes in the climate. The bumps and wiggles do, however, make determining trends over short time periods (e.g., a decade) problematic. To illustrate this, Figure 2.5 shows monthly average global

[1] www.foxnews.com/science/2013/09/27/un-climate-change-report-dismisses-slowdown-in-global-warming/, accessed 11/15/14.

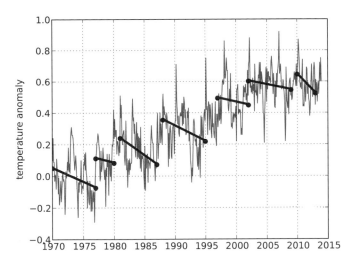

**Figure 2.5**  A plot of monthly and global average surface temperature from the surface thermometer record (gray line) along with short-term trend lines (black lines). This figure is an adaptation of SkepticalScience's escalator plot (www.skepticalscience.com/graphics.php?g=47 and www.skepticalscience.com/still-going-down-the-up-escalator.html)

surface temperature anomalies between 1970 and 2013. Over this period, the planet warmed rapidly, at a rate of 1.7°C/century.

Also shown on Figure 2.5 are short-term trends based on endpoints that were carefully selected to produce cooling trends. As you can see, it is possible to generate a continuous set of short-term cooling trends, even as the climate is experiencing a long-term warming. All you have to do is start the trend calculation during a particularly hot year (e.g., an El Niño year) and then end it in a cool year (e.g., a La Niña or volcanic year).

The existence of these short-term negative trends allows someone to disingenuously claim during almost any year covered by Figure 2.5 that global warming had stopped or even that the Earth had entered a cooling period. There is even a term for this deceptive argument: "Going down the up escalator."

There are two aspects of this worth emphasizing. First, claiming that global warming has stopped requires careful selection of the endpoints. This process of intentionally selecting data to yield a result counter to the full data set is known as *cherry picking*. Many of the skeptical claims you will hear in the public debate over climate are based on cherry picking a large data set in order to find the small number of exceptions that support the claim. Second, it is only possible to find cooling over short time periods. Over periods lasting several decades, the long-term warming dominates and even the most egregious endpoint selection cannot generate a cooling trend. We will return to this point when we talk about climate predictability in Section 8.7.

So two independent, direct measurements of temperature show the planet is warming, but this question is so important that even more confirmation is required. To do this, we turn to other measurements that, while not direct measurements of

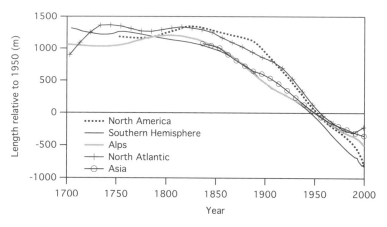

Figure 2.6 Change in mean glacier length over time, measured relative to length in 1950, for five world regions (the source is Figure 3.2 of Dessler and Parson, 2010, which was based on Figure 4.13 of Lemke et al., 2007).

temperature, nevertheless tell us something about the temperature of the planet: the amount of ice on the planet, the heat content of the ocean, and sea level.

## 2.1.3 Ice

Because ice melts reliably at 0°C, it is a dependable indicator of temperature. In particular, if the warming trend identified in the surface thermometer and satellite records is correct, then we should expect to observe the Earth's ice disappearing. In this section, I show that ice is indeed disappearing, thus confirming the warming seen in the other data sets.

## 2.1.3.1 Glaciers

Glaciers form in cold regions when snow that falls during the winter does not completely melt during the subsequent summer. As snow accumulates over millennia, the snow at the bottom is compacted by the weight of the overlying snow and turns into ice. This process eventually produces glaciers hundreds, or even thousands, of feet thick.

The length, areal extent, and total volume of glaciers have been monitored for decades and, in some cases, centuries. For example, Figure 2.6 shows changes in average glacier length (relative to the length in 1950) for five world regions over the past few centuries. It shows that glaciers began retreating around 1800, with the recession accelerating later in the nineteenth century. The pattern of glacier retreat is consistent worldwide, confirming that the warming we are now experiencing is truly global.

Decreases in precipitation or decreases in cloudiness could also cause glaciers to recede. However, the fact that glaciers are receding all over the planet means that, whatever is causing the changes, it must be global. And there is no evidence of global trends in either cloudiness or precipitation that could cause the reduction in glacier lengths. We do, however, have evidence of global trends in temperature

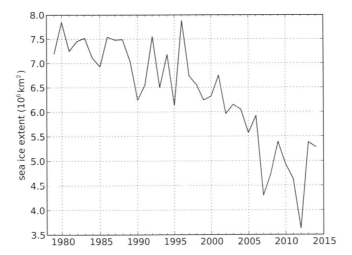

Figure 2.7 Arctic sea-ice area in September of each year (data obtained from the National Snow and Ice Data Center; see nsidc.org/data/seaice_index/).

(e.g., Figure 2.3). Thus, the recession of glaciers is consistent with the global warming of the climate seen in the surface thermometer and satellite records.

### 2.1.3.2 Sea ice

At the cold temperatures found in polar regions, seawater freezes to form a layer of ice floating on top of the ocean, typically a few meters thick. The area covered by sea ice varies over the year, reaching a maximum in late winter and a minimum in late summer. Given the rapid warming now occurring, particularly in the Arctic, we would expect to see reductions in the area covered by sea ice during the summer (during the winter, the temperatures are so low that a few degrees of warming does not have much of an effect). Figure 2.7 confirms this by showing a clear downward trend in the area covered by Arctic sea ice at the end of the summer. Measurements also show that, in addition to shrinking in area, sea ice has grown thinner.

The Antarctic is a different story. The sea-ice area around that continent has remained stable since the mid-1970s. This overall pattern – large losses of sea ice in the Arctic but little loss in the Antarctic – matches the regional temperature trends in these regions (Figure 2.2b), which shows large, rapid warming in the northern hemisphere and weaker warming in the southern hemisphere. In this way, the sea-ice data confirm not just the overall warming trend but also the global distribution of the warming.

### 2.1.3.3 Ice sheets

The Earth has two major ice sheets, one in the northern hemisphere, on Greenland, and the other in the southern hemisphere, on Antarctica. Although these ice sheets are really just big glaciers, their sheer size puts them in a class by themselves. They

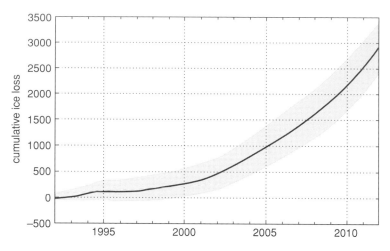

**Figure 2.8**  Cumulative loss of ice from the Greenland ice sheet, in billions of tons of ice. The shaded region is the uncertainty (adapted from Figure 4.15 of Vaughan et al., 2013).

contain the vast majority of the world's fresh water and, if they melted completely, the sea level would rise approximately 65 m. These ice sheets cover millions of square kilometers, and in places they are more than 3,000 m thick. Figure 2.8 shows the amount of ice lost between 1992 and 2013 from the Greenland ice sheet. Over this time, Greenland lost roughly 3,000 billion tons of ice, with the rate accelerating over time. Measurements from Antarctica show comparable losses for that ice sheet.

Putting all the data together, we can conclude that the amount of ice on the planet is decreasing. This is consistent with measurements of rising temperatures from surface thermometers and satellites and provides additional evidence that the Earth is warming.

## 2.1.4 Ocean temperatures

Much of the heat trapped by greenhouse gases goes into heating the oceans, so we can also look to see if the temperatures of the oceans are increasing. I am not talking about the surface temperature of the ocean – that is included in the surface thermometer record already described. Rather, I am talking about the temperature of the bulk of the ocean: the water temperature averaged over the entire depth of the ocean (the average depth of the ocean is 4 km). Scientists determine this temperature by lowering thermometers into the ocean and measuring the temperature at various depths and then averaging these results to come up with a single average ocean temperature over that depth.

Such measurements have been made for several decades, allowing us to analyze the temperature of the bulk of the ocean since the middle of the twentieth century; Figure 2.9 plots the time series of the ocean's temperature anomaly. The ocean is indeed observed to be warming, and this provides another source of independent confirmation that the Earth is warming. While the amount of warming of the ocean appears to be small, water holds a tremendous amount of energy, so this seemingly

**Figure 2.9**  Ocean temperature anomaly in degrees Celsius of the entire ocean. Anomalies are calculated relative to the 1970–2000 period (data are from Balmaseda et al., 2013).

small increase actually represents an enormous accumulation of energy in the climate system.

## 2.1.5 Sea level

Sea-level change is connected to climate change in two ways. First, as grounded ice[2] melts, the melt water runs into the ocean, increasing the total amount of water in the ocean and therefore the sea level. We saw in Subsection 2.1.3 that we are losing grounded ice all over the planet. Second, like most things, water expands when it warms, and we saw in Subsection 2.1.4 that the oceans are indeed warming, and the resulting expansion also raises sea level.

Figure 2.10 shows that is indeed what is observed. Over the period 1901–2010, global mean sea level rose by 0.19 m based on tide gauge records and satellite data since 1993. This corresponds to an average rate of sea level rise of 1.7 mm yr$^{-1}$. Between 1993 and 2010, the rate was higher: 3.2 mm yr$^{-1}$. Thus, sea level is not just rising, the rate of increase is accelerating.

Scientists can double-check these values by comparing them to estimates of the amount of water lost from grounded ice added to the amount of sea level rise due to warming of the ocean. Although the calculation is difficult, and there are uncertainties in the measurements (e.g., the warming of the deepest part of the ocean is not well constrained by observations), the loss of ice and the warming of the ocean is consistent with the observed change in sea level. Such detailed comparisons increase our confidence that the changes in each data set are accurate.

---

[2]  Grounded ice is ice that is resting on land. When it melts and the water runs into the ocean, sea level rises. This is different from floating ice. When floating ice melts, the melt water occupies roughly the same volume as was displaced by the ice (as shown by Archimedes), so there is little sea level rise. Thus, melting glaciers and ice sheets cause sea level to rise, but melting sea ice causes nearly none. I say "nearly" because sea ice is pure water, while seawater has salt in it; the differences in density between pure water and salt water leads to a small amount of sea level change when sea ice melts.

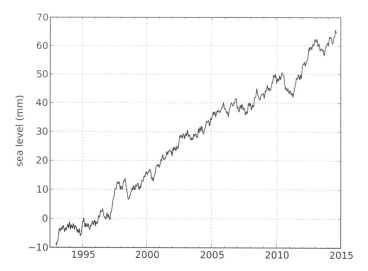

**Figure 2.10** Global-average sea level measured from satellites. The seasonal cycle has been removed. Data are described by Nerem et al. (2010) and were downloaded from sealevel.colorado.edu/content/2013rel7-global-mean-sea-level-time-series-seasonal-signals-removed.

### 2.1.6 Putting it all together: Is today's climate changing?

The answer is an emphatic *yes!* In fact, the evidence is so strong that the IPCC calls today's warming "unequivocal" – meaning it is beyond doubt. It is worth exploring the source of high confidence in this conclusion. First, there is great consistency among the various data sets. The surface thermometer record and the satellite record agree well, and both show that temperatures are rising. The loss of ice on the Earth's surface is consistent with these increasing temperatures, as is the increase in the heat content of the ocean. Finally, the observation of increasing sea level fits with both the loss of ice and the increasing heat content of the ocean.

These data sets are fundamentally independent. For example, issues such as changes in the station environment, which may affect the surface thermometer record, do not affect the satellite record. Issues such as orbit drift affect the satellite record but do not affect the surface thermometer record. And neither of these problems affects the measurements of glacier length or sea level. This means that there is no single problem or error that could introduce a spurious warming trend in all of the data. Because of this, there is virtually no chance that enough of these data sets could be wrong by far enough, and all in the same direction, that the overall conclusion that the climate is currently warming is wrong.

Moreover, the data sources we have reviewed are just a small part of the mountain of evidence that the Earth is warming. Other corroborating evidence includes decreased northern hemisphere snow cover, thawing of Arctic permafrost, strengthening of mid-latitude westerly winds, fewer extreme cold events and more extreme hot events, increased extreme precipitation events, shorter winter ice seasons on lakes, and thousands of observed biological and ecological changes that are consistent with warming (e.g., poleward expansion of species ranges and earlier spring flowering and insect emergence). Not every data set shows warming, but such contrary data are rare, regionally limited, and vastly outnumbered by evidence of warming.

### 2.1.7 What is *not* evidence of climate change

It is useful at this point to recognize what is *not* evidence of climate change. Because climate change is a shift in the statistics of the atmosphere, a single seemingly odd weather event is almost never evidence of climate change. A single extremely hot summer, for example, even if it were hotter than any other summer of the past 100 years, might nonetheless occur in a stable climate. If hot summers were to begin to happen regularly, however, then that would be indicative of climate change.

It is also important to avoid drawing conclusions about the global climate from regional climate extremes. Figure 2.1, for example, shows that temperatures in South America were much colder than average in June 2009. People living there might be forgiven for wondering where global warming was, but if they concluded as a result of those regional anomalies that climate change was no longer happening, they would be wrong. In that case, other regions were hotter than average, and those compensated for the low temperatures found in South America.

So be careful when evaluating the evidence for and against climate change. Do not be misled by unlikely-seeming single events or by regional occurrences. Neither is indicative of a shift in the global climate.

## 2.2 Climate over the Earth's history

### 2.2.1 Paleoproxies

To put today's warming into context, it is useful to consider the Earth's entire climate history. The measurements described in the previous section, however, go back at most a few centuries, so other data sets are needed if we wish to look back any further. Such data sets are known as *paleoproxies*, which are long-lived, geological, chemical, or biological systems that have the climate imprinted on them. In this way, we can make measurements *today* that tell us what the climate was like *in the past*.

For example, the ice in a glacier or ice sheet provides useful climate data dating back to the time when the snow fell. Remember that glaciers and ice sheets form when snow accumulates from one year to the next and is converted to ice by the weight of the overlying snow. The chemical composition of the ice holds important information about the air temperature around the glacier when the snow fell, as do variations in the size and orientation of the ice crystals. Small air bubbles trapped during the formation of glacial ice preserve a snapshot of the chemical composition of the atmosphere when the snow fell. In addition, the dust trapped in the ice gives information about prevailing wind speed and direction. And because more dust blows around during droughts, it also provides information about how wet or dry the regional climate was when the ice formed. Finally, because sulfur is one of the main effluents of volcanoes, measurements of sulfur in glacial ice show whether there was a major volcanic eruption around the time the ice formed.

To obtain all of this information, *ice cores* are obtained by drilling down into the glacier or ice sheet with a hollow drill bit and extracting a cylinder of ice a few inches in diameter. Reconstructing past climate information from an ice core then requires two steps. First, the age of each ice layer must be determined from its depth inside the glacier. The deeper down the ice was obtained, the older the ice is and the further back in time for which it provides climate information. Much effort has been spent connecting a particular chunk of ice to an exact time, because the rate of ice accumulation varies over time and because ice inside a glacier can compress and flow under the weight of the ice above. Second, the characteristics actually observed, such as the abundance of chemicals in the ice, must be translated into the climatic characteristics of interest, such as temperature.

Obviously, ice cores only provide climatic information in regions and over time periods that are cold enough for permanent ice to exist. This includes Greenland, Antarctica, and glaciers found around the world. Ice cores from the thickest, oldest ice in Antarctica have provided climate reconstructions dating back an amazing 800,000 years.

However, there are other paleoproxies that provide data in other regions and over other time periods. For example, trees also store climate information in their tree rings. Tree growth follows an annual cycle, which is imprinted in the rings in their trunks. As trees grow rapidly in the spring, they produce light-colored wood; as their growth slows in the fall, they produce dark wood. Because trees grow more, and produce wider rings, in warm and wet years, the width of each ring gives information about climate conditions around that tree in that year. So by looking at the rings of a tree, scientists obtain an estimate of the local climate around the tree for each year during which the tree was alive.

Climate data from tree rings are only available for a fraction of the Earth's surface. They are obviously not available for oceans, or for desert or mountainous areas where no trees grow. They are also not available in the tropics, where the weaker seasonal cycle causes trees to grow year round; those trees do not produce rings. Finally, tree rings only reveal information about the climate as far back as trees are available. This means that the longest tree ring records provide useful climate information for the past millennium or so.

Ocean sediments, which accumulate at the bottom of the ocean every year, also contain information about climate conditions at the time they were deposited. The most important source of information in sediments comes from the skeletons of tiny marine organisms. The relative abundance of species that thrive in warmer versus colder waters gives information about surface water temperature. The chemical composition of the skeletons and variations in the size and shape of particular species provide additional clues. In total, ocean sediments provide information about water temperature, salinity, dissolved oxygen, atmospheric carbon dioxide, nearby continental precipitation, the strength and direction of the prevailing winds, and nutrient availability; this information goes back tens of millions of years.

Putting all of these paleoproxies together gives us a reasonably complete picture of the global climate going back many millions of years, with some information about the climate going back billions of years.

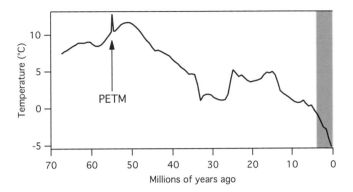

**Figure 2.11** Reconstructed temperature of the polar regions over the past 70 million years. The sharp temperature spike 55 million years ago represents the PETM. The gray bar on the right shows the past 4 million years, which are expanded in Figure 2.12. Because this time series is from ocean isotopes, it is sensitive both to temperature and to the total volume of ice on land. Starting roughly 35 million years ago, some of the variation here comes from changes in land ice volume rather than temperature. The overall trend, however, mostly represents changes in temperature (the source is Figure 3.8 of Dessler and Parson, 2010, which was adapted from Figure 2 of Zachos et al., 2001).

## 2.2.2 The Earth's long-term climate record

Although many of details of the climate during the first 97 percent of the Earth's history are unknown, there are a few things that we can say. To begin with, the oldest sedimentary rocks on the planet are nearly 4 billion years old. Because sedimentary rocks generally form in the presence of liquid water, their existence suggests that the Earth has been warm enough over most of its history that water has remained mostly in the liquid phase.

While the Earth has generally been warm, there is also evidence of intervals of widespread ice cover (known as a glaciations). The evidence comes in the form of marks on rocks, such as abraded rock surfaces and other geologic formations that form when giant ice sheets flow over rocks. It shows that, approximately 700 million years ago, the Earth was covered by ice from the poles to near the equator – a climate configuration now referred to as *snowball Earth*. There was also a significant planetary glaciation roughly 300 million years ago.

Figure 2.11 shows a reconstruction derived from ocean sediments of polar temperatures over the past 70 million years. The warmest temperatures in this record occurred approximately 50 million years ago – 15 million years after the extinction of the dinosaurs – in a period called the *Eocene Climatic Optimum*. During that time, the planet was far warmer than it is today. Forests covered the Earth from pole to pole, and plants that cannot tolerate even occasional freezing lived in the Arctic, along with animals such as alligators that today live only in tropical climates. Since that time, the Earth has experienced a long-term cooling. Clearly, humans had nothing to do with either the warmth of the Eocene or the cooling since then; I will talk more about the causes of these climate variations in Chapter 7.

Figure 2.11 also shows the Paleocene-Eocene Thermal Maximum (frequently abbreviated PETM), which occurred 55 million years ago, at the temporal boundary

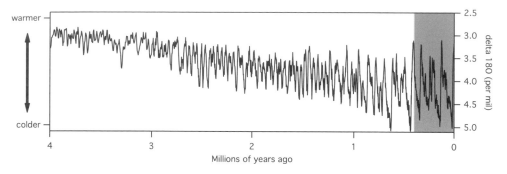

**Figure 2.12**   Measurement of global average relative temperature over the past 4 million years. The vertical axis measures the relative abundance of oxygen-18, the heavy isotope of oxygen that is a proxy for temperature, in ocean sediment cores. The temperature difference between the top and bottom of the graph is roughly 10 °C. The gray bar on the right shows the past 410,000 years, which are expanded in Figure 2.13 (the source is Figure 3.9 of Dessler and Parson, 2010, which is based on the analysis of Lisiecki and Raymo, 2005).

between the Paleocene and Eocene epochs. The PETM featured an abrupt warming of a few degrees Celsius that occurred over a few thousand years. The temperature then slowly returned back to pre-PETM temperatures over the next few hundred thousand years. It is believed that this was caused by a massive release of greenhouse gases; many scientists view this episode as a good analog to the warming event we are now in the midst of. I will also discuss this event in more detail in Chapter 7.

Figure 2.12 zooms in to show global temperature variations over the past 4 million years. Like the 70-million year record in Figure 2.11, this record also shows a general cooling trend. This record, however, covering a more recent time, shows fine-scale details that are not visible in the longer record. For example, starting roughly 3 million years ago, around the time that large ice sheets first appeared in the northern hemisphere, large oscillations between warmer and cooler periods suddenly appear in the record. During the cool periods, called *ice ages*, the ice sheets expanded to cover large parts of the northern hemisphere's land areas. During the warm periods between the ice ages, called *interglacials*, the ice sheets contracted. From approximately 2.5 million to 1 million years ago, ice ages occurred every 41,000 years. Since then, for reasons that are not well understood, the frequency of ice ages shifted to every 100,000 years and the magnitude of the ice-age cycles increased.

Figure 2.13 zooms in again, showing a record of temperature and carbon dioxide levels for the Antarctic region over the past 410,000 years constructed from ice cores. This record shows in more detail the shape of ice-age cycles – the cooling into an ice age is slow, taking several tens of thousands of years, whereas the warming at the end of an ice age occurs faster, in approximately 10,000 years. Overall, these ice ages lasted about 100,000 years, while the interglacials are relatively short, lasting 10,000 to 30,000 years. The last ice age ended roughly 12,000 years ago, and since then the Earth has been enjoying a rather pleasant interglacial.

It is particularly noteworthy that atmospheric carbon dioxide, which can also be estimated from the ice core, varies closely with atmospheric temperature over these ice age cycles. I will discuss the implications of this relationship in detail in

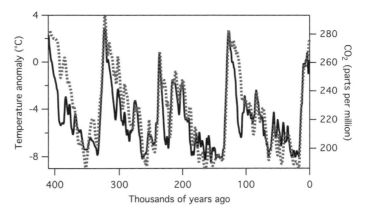

**Figure 2.13** Temperature anomaly of the southern polar region (solid line) over the past 410,000 years, relative to today's temperature, constructed from an Antarctic ice core. Carbon dioxide (dotted line) is from air bubbles trapped in the ice (the source is Figure 3.10 of Dessler and Parson, 2010, which was adapted from Petit et al., 1999).

Chapter 7, but you would be correct if you concluded that carbon dioxide variations play a key role in the generation of the ice-age cycles.

Figure 2.14 zooms in again to show the global temperature of the Holocene, the period beginning at the end of the last ice age, about 11,700 years ago. This estimate shows that temperatures peaked about 7,000 years ago and then started a slow, long-term decline that bottomed out in a period 200 to 300 years ago, known as the *Little Ice Age*, during which temperatures were about 1 °C below today's. After the Little Ice Age, temperatures began warming. Today's temperatures are certainly warmer than most of the Holocene; however, given all of the uncertainties, it is unclear whether the mid-Holocene was warmer than today.

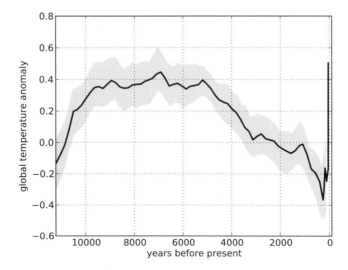

**Figure 2.14** Global temperature anomaly of the last 11,000 years, based on multiple proxy records. The shaded represents the uncertainty in the estimate. Anomalies are calculated relative to the 1961–1990 average (the source is Figure 1B of Marcott et al., 2013).

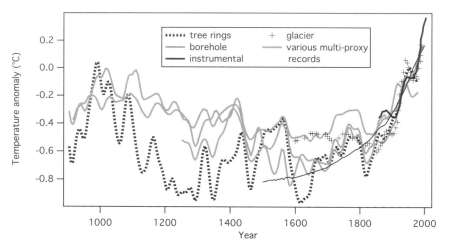

Figure 2.15 Average northern hemisphere temperature anomaly over the past 1,000 years, based on multiple proxy records and the modern surface thermometer record. Anomalies are calculated relative to the 1961–1990 average (the source is Figure 3.11 of Dessler and Parson, 2010, which was adapted from Figure S-1 of the National Research Council, 2006).

---

**An aside: What does the paleorecord tell us about how serious of a threat climate change is?**

As we will talk about in Chapter 8, forecasts for the twenty-first century are for a few degrees Celsius of warming. That might not seem like much, but the paleorecord says otherwise. The ice ages were only about 5°C colder than today, and the Earth was essentially a different planet. Glaciers several thousand feet thick covered much of North America, sea level was 100 m lower than today, and there were significant accompanying changes in the world's environment and ecosystems.

The Little Ice Age was only about 1°C below today's temperature, a seemingly trivial amount, but the climate was different enough that we call it the Little Ice Age. Glaciers in Europe advanced dramatically, destroying numerous farms and villages. Paintings from the time show a cold and snowy climate that does not exist today. In London, the freezing of the Thames River, commonplace during the era, was celebrated with a winter fair that took place on the frozen river – that river no longer freezes over. In their camp in Valley Forge, PA, the Continental Army nearly froze to death during the winter of 1777–1778.

Thus, we should expect warming of a few degrees Celsius, if it comes to pass, to radically change the planet we're living on.

---

Figure 2.15 zooms in one last time to show average northern hemisphere temperature over the past 1,000 years, based on multiple proxies and modern records. Once again, this figure shows short time-scale temperature variations that are not visible in the graphs covering longer time spans. The various estimates differ, particularly before Year 1500, but all show a similar pattern. Temperatures were warm 1,000 years ago, during a period known as the *Medieval Warm Period*. There were then several centuries of gradual cooling, bottoming out in the Little Ice Age, followed by faster warming since the nineteenth century.

This vast and growing body of information about the Earth's past allows us to make several important conclusions. We can say with high confidence that the past few decades of the twentieth century are warmer than any comparable period during the past 400 years, and possibly even warmer than the peak of the Medieval Warm Period, around 1,000 years ago. It is presently unclear whether today's temperatures exceed the previous peak temperature of the Holocene, which occurred about 7,000 years ago.

Over geologic time scales (millions of years or longer), we can say with high confidence that the Earth has been both far warmer and far cooler than it is today. But this fact provides no information about the cause of today's warming. As I will discuss in Chapter 7, there is strong evidence to suggest that the recent warming is primarily due to human activities.

We can also say that the warming of the past few decades has been rapid. For example, the warming over the past century (approximately 0.8°C in 100 years) is occurring at least ten times faster than the average rate of warming coming out of the last ice age (roughly 5°C in 10,000 years corresponds to an average warming of 0.05°C/century). This means that projections of warming of several degrees Celsius over the next century would be both large and exceptionally rapid.

The challenge for the scientific community is to come up with a theory that explains all of the variations in the climate record, from snowball Earth 700 million years ago to the rapid warming of the past few decades. In the next few chapters, we will learn about the fundamental physics that governs our climate, and then, in Chapter 7, we will put it all together to show that most of the warming of the past few decades can be attributed to human activity.

## 2.3 Chapter summary

- The most well-studied and reliable source of temperature data for the past century is the surface thermometer network. It shows a global and annual average warming of 0.85°C between 1880 and 2012, with an uncertainty of ±0.2°C.
- Scientists have a large number of independent measurements with which to confirm the warming seen by the surface thermometer network. These include satellite measurements of temperature, measurements of the amount of ice on the planet, ocean heat measurements, and sea-level measurements. All of these data confirm the warming seen in the surface thermometer data.
- Because of the overwhelming evidence supporting it, the scientific community has concluded that the observed warming of the climate system is beyond doubt – the IPCC uses the word "unequivocal." Furthermore, scientists have concluded that the previous decade is very likely the hottest of the past 400 years.
- Looking back further in time, we see that the Earth's climate has varied widely over its 4.5-billion-year history. The geologic record shows that the climate has been both warmer and cooler than today's climate.

- Over the past few million years, the Earth has oscillated between ice ages and warmer interglacial periods. Ice ages are about 5°C cooler than the interglacials. The Earth is currently in an interglacial.

## Additional reading

The Working Group I report of the IPCC's Fifth Assessment describes, at varying levels of detail, the evidence supporting the conclusion that warming is "unequivocal." For the most detail, see Chapter 2 of the report. For a less detailed overview, see Section TS.2 of the Technical Summary. And, for a short, high-level discussion, see Section B of the IPCC's Summary for Policymakers (you can download all of these at www.ipcc.ch/report/ar5/wg1/).

One of the internet's great resources for climate information is Skeptical-Science.com. It has many articles about the science of climate change, including discussions of the quality of the various temperature records. For example, here is a nice article on the reliability of the surface thermometer measurements: www.skepticalscience.com/surface-temperature-measurements.htm. When you are confronted with a claim that sounds wrong ("We're entering a new ice age!"), this is the first place you should go to check it out.

See www.andrewdessler.com/chapter2 for additional resources for this chapter.

## Terms

Cherry picking
Eocene Climatic Optimum
Error bar
Ice ages
Ice core
Interglacials
Little Ice Age
Medieval Warm Period
Paleoproxies
Snowball Earth
Temperature anomaly

## Problems

1. Every year, you measure the height of a child relative to a coat hook on the wall. In the first year, he was 2" below the hook, the next year he was 1.5" below the hook, the next year he was 0.75" below the hook, the next year he was even with

the hook, the next year he was 0.5" above the hook, and the last year he was 1.5" above the hook. (a) What was the total amount the child grew? (b) What was his average growth rate (in inches per year)? (c) What was his absolute height at the end of the last year?

2. How much did the Earth warm between a) 1880 and 2012 and b) 1970 and 2012? Provide answers in both degrees Celsius and degrees Fahrenheit.

3. If you found out that the satellite data were unreliable because of a previously unknown error, would that change your opinion about whether the Earth is currently warming? Why or why not?

4. A reporter asks you to explain why scientists are so confident that the Earth has undergone a general warming over the past few decades. Knowing that reporters hate long answers, construct an answer that takes 60 seconds or less to deliver.

5. List the evidence that supports the contention that the Earth is currently warming. Is there any evidence that goes against this conclusion?

6. What is a temperature anomaly? Why are temperature anomalies typically used in global temperature calculations?

7. Download the annual and global average temperature data from the NASA GISS (Google the term *GISTEMP*) and reproduce Figure 2.2. Calculate your own trends for the past thirty years and for the past 100 years.

8. For the more adventurous students: Download the monthly and global average temperature data from the NASA GISS (Google the term *GISTEMP*). Calculate trends over various lengths of time (ranging from a decade to several decades). Over what length of time can you frequently find negative trends; over what length of time are the trends mainly positive?

9. Why do we turn to paleoproxy measurements to infer the temperature of millions of years ago?

10. Go to cdiac.ornl.gov/epubs/ndp/ushcn/access.html and plot up the monthly temperature for the past 100 years for the station nearest your hometown. Does this look like the global average time series in Figure 2.2? Should it?

11. A global warming advocate tells you that the Earth is now warmer than it has ever been. Is that correct?

# Radiation and energy balance

The Earth's climate is a complex physical system. Nevertheless, we can still understand much about the climate even without an advanced degree in physics. In this chapter, I introduce the important physics required to understand the climate. Then, in Chapter 4, we will use this physics to construct a simple model of our climate.

## 3.1 Temperature and energy

Before we get into the physics of climate, it is useful to first talk about the concept of *energy*. To a physicist, energy is the capacity to do work – such as lifting a weight, turning a wheel, or compressing a spring. The unit of energy most frequently used in physics is the *joule,* abbreviated as the letter J, and 1 J is approximately the amount of energy required to lift 100 g about 1 m – or to lift an apple about 3 ft.

Energy often moves from one place to another. The rate at which energy is flowing is referred to as *power*. It is usually expressed in *watts,* abbreviated as the letter $W$. One watt is equal to one joule per second – that is, $1\ W = 1\ J/s$ – so a 60-W light bulb consumes 60 J of energy every second.

An analogy may help to illuminate the difference between power and energy. A gallon is a quantity, such as a gallon of water. This is akin to a joule, which is a quantity of energy. The rate at which water flows through a pipe is measured in, say, gallons per minute. The rate at which energy flows is the power, and it is measured in watts (joules per second). As you will see in this chapter and the next, climate is all about energy flows, so climate calculations mostly focus on power.

---

An example: How much power does it take to run a human body?

A typical human consumes approximately 2,000 food calories per day. Calories are an alternative unit of energy, where 1 food calorie = 4,184 J. Thus, 2,000 food calories corresponds to 8,368,000 J = 8.37 MJ. One day has 24 hours × 60 minutes × 60 seconds = 86,400 seconds in it, so dividing 8,368,000 J by 86,400 s yields 97 J/s = 97 W. Thus, the typical human requires roughly 100 W to power his or her body – about the same power required to run a light bulb or two. One horsepower is approximately 740 W, so another way to think about this is that it takes about one seventh of a horsepower to run your body.

---

The *internal energy* of an object refers to how fast the atoms and molecules in the object are moving. In a cup of water, for example, if the water molecules are moving slowly, then the water has less internal energy than another cup in which the molecules of water are moving rapidly. In a solid, the movements of the atoms are approximately fixed in space by intermolecular forces – that is why it is a solid. The atoms, however, can still move small distances around their fixed position. The faster these atoms move about their fixed position, the more internal energy the object has.

This brings us to a concept that most people are familiar with: *temperature.* Temperature is a measure of the internal energy of an object. As an object's internal energy increases and the molecules of the object speed up, the temperature of the object also increases. Thus, if you have two cups of water, one hot and the other cold, you can conclude that the water molecules in the hot cup are moving faster than the water molecules in the cup of cold water.

In Chapter 1, I introduced the Celsius temperature scale, which scientists use frequently. There is another temperature that is even more favored by physicists, and it is called the *Kelvin scale.* The temperature in Kelvin is equal to the temperature in degrees Celsius plus 273.15 $(K = C + 273.15)$. Thus, the freezing temperature, $0\,°C$, is equal to 273.15 K, whereas the boiling temperature, $100\,°C$, is equal to 373.15 K. "Room temperature" is $22\,°C$ or so, which is 295 K. Most temperatures found in the Earth's atmosphere are between 200 K and 300 K, and the average surface temperature of the Earth (today, at least) is roughly 288 K.

Physicists prefer the Kelvin scale because temperature expressed in Kelvin is proportional to internal energy. Thus, if the temperature doubles from 200 K to 400 K, then the internal energy of the object also doubles. If the internal energy of an object increases by 10 percent, then the temperature expressed in Kelvin also increases by 10 percent. And 0 K is absolute zero – the temperature at which molecules have zero internal energy and cease moving[1]; this is the coldest possible temperature. Because of this important quality, the physics equations introduced in this chapter and the next require energy to be expressed in Kelvin.

## 3.2 Electromagnetic radiation

It has long been recognized that the warmth of our climate is provided by the Sun. However, the Sun sits 150 million km away from the Earth, with the vacuum of space in between. How does energy from the Sun reach the Earth?

Energy is transported from the Sun to the Earth by *electromagnetic radiation.*[2] Electromagnetic radiation includes visible light, like that put out by your desk lamp

---

[1] Note that you do not use a degree sign (°) when writing temperature in Kelvin. Thus, room temperature is 295 K, not 295°K. That is because Kelvin is an absolute temperature scale and 0 K is absolute zero.

[2] When people hear the word *radiation,* they often think of nuclear radiation. Such radiation has very high energies because it originates from changes in atomic nuclei, and as a result this radiation can cause cancer and other medical problems. Electromagnetic radiation discussed here generally originates from changes in the atoms' electrons or from changes in the molecule's rotational or vibrational state

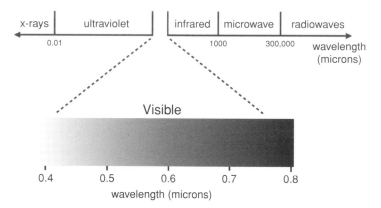

Figure 3.1 The electromagnetic spectrum. Note that the visible part makes up only a minor part of this spectrum. (See Color Plate 3.1.)

or the Sun, X-rays, like those that allow us to detect broken bones, microwaves, like those that cook your dinner, and radio-frequency waves, like those that bring calls to your cell phone and WiFi to your computer.

One can think of electromagnetic radiation as a stream of *photons*, small discrete packages of energy.[3] As photons travel from Point *A* to Point *B* – such as from the Sun to the Earth – each one carries a small amount of energy, and this is how energy is transported from the Sun to the Earth.

Photons have a characteristic size, referred to as the *wavelength*, which determines how the photons interact with the world. Photons with wavelengths of between 0.3 and 0.8 microns (a *micron*, abbreviated as µm, is a millionth of a meter; a human hair is 100 µm or so in diameter) can be seen with the human eye – so we refer to these photons as *visible*. Within the visible range, the different wavelengths appear to the human eye as different colors (Figure 3.1). Humans see photons with wavelengths near 0.4 µm as blue, 0.6 µm as yellow, and 0.8 µm as red.

Photons with longer wavelengths, from 0.8 to 1,000 µm, are termed *infrared* – from the Latin for "below red" – because they are beyond the red end of the visible spectrum. Despite being invisible to humans, these photons play an important role in both the Earth's climate and in our everyday lives. Photons with wavelengths just below the human detection limit of 0.3 µm are called *ultraviolet* because their wavelength is beyond the violet end of the visible spectrum.

Photons with wavelengths between 1,000 µm (1 mm) and 0.3 m are termed *microwaves*, and photons in this wavelength range are used in many familiar applications, from cooking to radar. Wavelengths longer than about 0.3 m are radio-frequency waves, and they are used, as the name implies, in radio applications. The entire electromagnetic spectrum is diagrammed in Figure 3.1.

The wavelength determines a photon's physical properties. For example, visible and infrared photons cannot go through walls, but radio-frequency photons can. The

and therefore has far less energy – so it is generally not a health hazard. This is good, because you are surrounded by electromagnetic radiation right now.

[3] Electromagnetic radiation also behaves like a wave, but for this problem it is easier to think of it as a particle.

human eye can detect visible photons but not infrared or microwave photons. When you get a full body scan at the airport's security checkpoint, the machine is most likely using microwaves – that wavelength goes through clothes but is stopped by denser materials such as flesh, a bomb, or a gun. Finally, the atmosphere is transparent to visible photons but not to infrared photons; this fact has enormous implications for our climate and will be discussed at length in Chapter 4.

## 3.3 Blackbody radiation

We know that both the Sun and the lamp on your desk are emitting photons. After all, you can see the visible photons that they are emitting. They are not, however, the only things around you that are emitting photons. In fact, *everything around you* is emitting photons all of the time. So right now, you are emitting photons, as are the walls of the room you are sitting in, your desk, your dog, this book. Everything.

If everything is emitting photons, then why does not everything glow like a light bulb? It turns out that the wavelength emitted is determined by the object's temperature. Figure 3.2 plots *emissions spectra* for idealized objects called *blackbodies* at three temperatures. An emissions spectrum shows the power carried away from an object by the photons at each wavelength.

Figure 3.2a shows the distribution of photons emitted by a 300-K blackbody, about room temperature. Photons emitted by this object almost exclusively have wavelengths greater than 4 μm or so. These wavelengths are outside the range that is visible to humans (indicated by the gray shading in the figure). Thus, all room-temperature objects are emitting photons, but you cannot see the photons because they fall outside the visible range. This is, in fact, the origin of the term *blackbody*. At room temperature, the object appears black because the photons emitted by these objects are invisible to humans (also playing a role is that blackbodies absorb all photons that fall on them – they do not reflect any). Blackbodies are idealized constructs, but most objects nevertheless behave like one, at least approximately.

Figure 3.2a also shows that the peak of the emissions spectrum for a 300-K blackbody occurs near 10 μm. It turns out that there is a simple relation between the temperature and the peak of the object's emission spectrum. This relation is known as *Wien's displacement law*:

$$\lambda_{max} = 2897/T \tag{3.1}$$

$T$ is the temperature of the blackbody in Kelvin and $\lambda_{max}$ is the wavelength of the peak of the emission spectrum in microns. If we put 300 K into Equation 3.1, we get 9.7 μm, which is in good agreement with Figure 3.2. Note the importance of using Kelvin temperature – had I used the temperature in degrees Celsius, I would have calculated $\lambda_{max} = 2897/27 = 107$ μm, which would be incredibly wrong.

Wien's displacement law also tells us that, as an object heats up, $\lambda_{max}$ decreases, shifting the peak of its emission spectrum to shorter wavelengths. Thus, a 1,600-K object has $\lambda_{max} = 1.9$ μm and a 6,000-K object has $\lambda_{max} = 0.5$ μm, values consistent with the emissions spectra in Figures 3.2b and 3.2c.

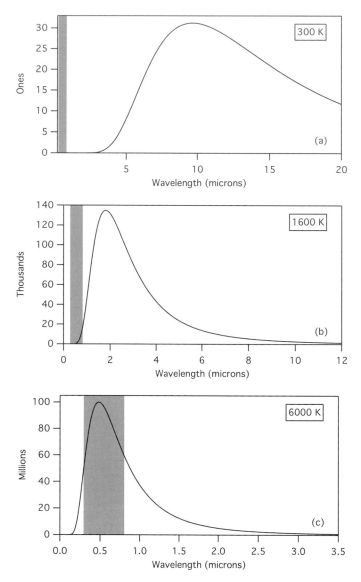

**Figure 3.2** Power emitted at different wavelengths from objects (with surface area of 1 m²) at three temperatures: (a) 300 K, (b) 1,600 K, and (c) 6,000 K. The vertical axes are in units of 1 W/µm, 1,000 W/µm, and 1 MW/µm of wavelength range, respectively. Gray bars show the wavelength range visible to human eyes.

It is worth emphasizing that objects do not just emit photons at $\lambda_{max}$; they emit them over a range of wavelengths around $\lambda_{max}$. So, while $\lambda_{max} = 1.9$ µm for the 1,600-K object, the object emits photons over a range of wavelengths from 0.7 to 10 µm. Because a fraction of the photons emitted by this object have wavelengths between 0.7 and 0.8 µm, which lie at the red end of the visible spectrum, humans will perceive a 1,600-K object as having a slight reddish glow. In other words, this object is glowing "red hot." Blacksmiths use this fact to determine when a piece of metal has reached an appropriate temperature, and the necessity of seeing a faint glow from an object is one reason that blacksmiths often work in dim, low-light conditions.

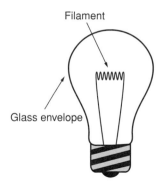

**Figure 3.3**   A schematic of a typical incandescent light bulb.

For the 6,000-K object, most of the photons emitted fall within the visible range (Figure 3.2c). Our Sun is, to a good approximation, a 6,000-K blackbody, and the distribution of photons from the Sun is closely approximated by this blackbody spectra. Because being able to see confers a strong advantage in surviving, it is no surprise that the eyes of humans and other animals have evolved to see this range of wavelengths. In fact, the human eye is maximally sensitive to light with a wavelength near 0.5 μm, which is the $\lambda_{max}$ for a 6,000-K blackbody. The chlorophyll molecule, the key component of photosynthesis, strongly absorbs photons in the visible range, showing that plants have also evolved to take advantage of photons emitted by the Sun.

Finally, if the photons emitted by room-temperature objects are not visible to our eyes, how can we see room-temperature objects, such as this page? What you see when you look at a room-temperature object are visible photons (emitted by the Sun or a light bulb or some other hot object) that have bounced off the object.

An everyday object that uses a lot of the concepts that we have discussed in this chapter is the humble *incandescent light bulb*. An incandescent light bulb consists of a glass envelope containing a small filament made of a metal, such as tungsten. When the light bulb is turned on, electricity flows through the filament, heating it to around 3,000 K (Figure 3.3).

Figure 3.4 shows the wavelength distribution of photons emitted by a 3,000-K blackbody. As the figure shows, the filament is hot enough that some of the photons

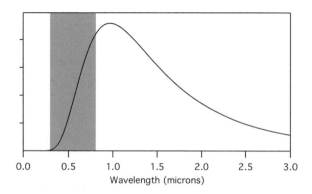

**Figure 3.4**   Emissions spectrum for a 3,000-K blackbody, a typical filament temperature for an incandescent light bulb. The numbers on the y-axis are omitted.

emitted are visible – so humans will see the light bulb glowing and you can use it to light your room. However, 85 percent of the photons emitted have wavelengths in the infrared, too long for the human eye to detect. These infrared photons provide no lighting for humans and so the energy to produce them is essentially wasted. This makes incandescent bulbs inefficient as light sources.

One way for a light bulb to produce a higher fraction of visible photons is to run the filament at a higher temperature. As described by Equation 3.1, this shifts the distribution of emitted photons to shorter wavelengths, thereby making a greater fraction of them visible to humans – thereby making the bulb more efficient. The optimal temperature for the filament would be about the temperature of our Sun, nearly 6,000 K, which provides the best overlap between blackbody emission and the human visual range. Unfortunately, at such a temperature, the filament would immediately vaporize and the bulb would be destroyed.

In fact, conventional incandescent bulbs are run at about as high a temperature as they can be. To further increase the filament temperature, the nitrogen and argon found in most bulbs is replaced with halogen gas. Because of chemical reactions between the halogen gas and the filament, the filament in these so-called halogen light bulbs can survive at temperatures several hundred degrees hotter than a regular incandescent bulb. This means that halogen light bulbs put out more visible photons, making them more efficient than regular incandescent bulbs. Unfortunately, because the filament is run so hot, the light bulb itself also gets extremely hot, creating a fire and burn hazard.

A better way to obtain high efficiency lighting is to change the technology. Compact fluorescent light bulbs (CFL) and light emitting diode (LED) light bulbs use different technologies (which I will not discuss here) to emit most of the bulb's photons in the visible wavelength range. The net result is a bulb that is at least five times more efficient. In other words, a 12-W CFL will produce the same amount of visible light as a 60-W incandescent light bulb – and it does this by reducing the amount of infrared light emitted. In an effort to boost energy efficiency, the U.S. Congress passed a law in 2007 phasing out production of standard incandescent bulbs in the United States by 2014. By the time you read this, it may be difficult to find those bulbs being sold.

Not only does the wavelength of emission change with temperature, but the total power emitted also increases with temperature. The observant reader would have seen this in the varying y-axis ranges in Figure 3.2, but it is explicitly shown in Figure 3.5a, which shows four different blackbody-emission curves on a single plot. The plot shows that, at every wavelength, warmer objects emit more power than cooler objects.

For a different view of this, Figure 3.5b plots the total power emitted by a blackbody as a function of temperature. It is clear that, as the temperature of the object increases, so does the power emitted. It turns out that there is a simple relation, known as the Stefan-Boltzmann equation, between the total power radiated by a blackbody and temperature:

$$P/a = \sigma T^4 \tag{3.2}$$

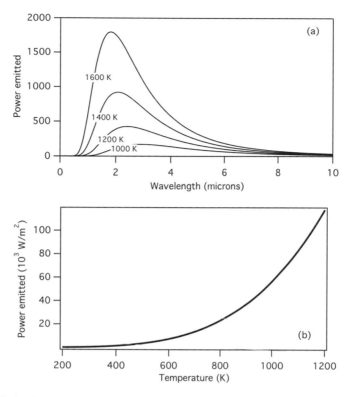

**Figure 3.5** Plots of (a) the distribution of power emitted by a blackbody at four different temperatures (1,600, 1,400, 1,200, and 1,000 K), in $(W/m^2)/\mu m$, and (b) total power emitted by a blackbody as a function of temperature, in $W/m^2$.

$P/a$ is the power emitted by a blackbody per unit of surface area, with units of watts per square meter; $\sigma$ is the Stefan-Boltzmann constant, $\sigma = 5.67 \times 10^{-8}$ $(W/m^2)/K^4$; and $T$ is the temperature of the object in Kelvin. If you multiply $P/a$ by the surface area $a$ of the object (in square meters), then you get the total power emitted by a blackbody, in watts.

The Stefan-Boltzmann equation has wide applications. By measuring the amount of power emitted by an object, astronomers use it to infer the temperature of distant stars and planets. The U.S. military uses the equation to build sensors to identify and lock onto hot jet engines against a cold sky in the guidance systems of heat-seeking missiles. And the ear thermometer that might be in your medicine cabinet right now uses this same physics to convert the infrared emission from the eardrum and surrounding tissue into an estimate of body temperature.

As a further example, Figure 3.6 shows an image of my dog, Kasper, in the infrared. To construct this image, the temperature is determined by measuring the infrared emission and converting this to temperature. Bright colors indicate warm temperatures and dark colors indicate cool temperatures. Like humans, dogs are mammals and their body temperature is around 38 °C. Fur is an insulator, however, so fur-covered regions of the dog tend to be cooler. Areas that are not fur covered, such as the eyes, are close to the dog's internal temperature. Note also the dog's cold nose.

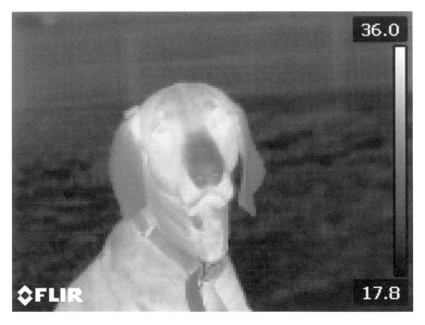

**Figure 3.6**    Photo of Kasper Dessler in the infrared, with colors assigned to different temperatures. (See Color Plate 3.6.)

---

An example: How fast is a room-temperature basketball losing energy by the emission of photons?

At room temperature, a blackbody is emitting $\sigma(300 \text{ K})^4 = 460 \text{ W/m}^2$. A basketball with a radius of 5 in. $= 0.13$ m has a surface area of $4\pi(0.13 \text{ m})^2 = 0.2 \text{ m}^2$. The total rate of energy loss from a room-temperature basketball as a result of blackbody photon emission is therefore $460 \text{ W/m}^2 \times 0.2 \text{ m}^2$, or 92 W. This is about the same emitted power as a typical light bulb. Of course, you cannot light a room with a basketball because the photons emitted by a basketball are outside the range that humans can see.

---

## 3.4 Energy balance

One of the cornerstones of modern physics is the first law of thermodynamics, which basically says that *energy is conserved.* If some object loses some energy, then some other object must gain that same amount of energy. Furthermore, because photons are just little packets of energy, the first law tells us that when an object emits a photon, the object's internal energy must decrease. And because temperature is a measure of internal energy, the emission of a photon therefore causes the object to cool. Similarly, if a photon hits an object and is absorbed, then the energy of the photon is transferred to the object's internal energy and the object warms.

An example: conservation of money

A good analogy for energy balance is money balance. If you gain one dollar, then someone else must be one dollar poorer because money, like energy, cannot be created or destroyed.[4]

Consider, for example, a checking account. Funds, such as your paycheck or a birthday check from your grandmother, are periodically deposited into the account. At the same time, funds are withdrawn, to pay for things such as rent or a cell phone bill. The change in your bank balance is equal to the difference between the total deposits (*money in*) and total withdrawals (*money out*). In equation form, we write this as follows:

$$\text{Change in balance} = \text{money in} - \text{money out}$$

If *money in* exceeds *money out*, that is, your deposits exceed your withdrawals, then the change in balance is positive and your balance increases. If *money out* exceeds *money in*, then the change in balance is negative and your balance decreases. If *money in* and *money out* are equal, the change in balance is zero and your balance is unchanged. This is basically the calculation we do when we balance our checkbooks.

If the energy flowing into an object (*energy in*) exceeds the energy flowing out (*energy out*), then the internal energy (and temperature) of the object increases. Written mathematically, this is as follows:

$$\text{Change in temperature } \alpha \text{ Change in internal energy} = \text{energy in} - \text{energy out}$$

Here the symbol $\alpha$ means "is proportional to." Note that, if *energy in* and *energy out* are equal, the internal energy and temperature are unchanged. We call this situation *equilibrium*.

A good example that draws many of the concepts in this chapter together is your home oven. Most people, if asked how an oven cooks, would answer, "Because it's hot inside." However, you may be surprised that the physics is subtler than you realize. Ovens do not cook because the air in the oven is hot – air is a terrible conductor of heat. Rather, ovens cook by infrared radiation.

When an electric oven is turned on, electricity runs through a heating element. The element heats up, eventually reaching temperatures high enough that it glows a dark orange. At this point, the element is radiating an enormous amount of power, typically several thousand watts.

The photons emitted by the heating element are absorbed by the walls of the oven, heating them. When the walls reach a predetermined temperature, typically 350–450°F (450–500 K), then the oven is "preheated" and the cook puts the food, say a turkey, into the oven. Let us assume the turkey came out of the refrigerator and has a temperature of 3°C or 276 K. At this temperature, the turkey is radiating 330 W/m$^2$. If the turkey has a surface area of 0.1 m$^2$, then the total power radiated by the turkey is 33 W.

---

[4] This rule does not apply to national governments, which can print money.

The turkey is also absorbing photons from the oven's hot walls. The oven walls, at 375°F (465 K), are radiating 2,650 W/m². The total surface area of the oven's six walls is approximately 1.3 m², so the total power radiated by the oven's wall is roughly 3,500 W. Most of the energy radiated by the oven's walls misses the turkey in the middle and hits the other walls, and only a fraction of photons emitted by the walls hits the turkey. The turkey absorbs photons emitted by an area of the walls equal to the surface area of the turkey, 0.1 m². Given that the walls emit 2,650 W/m², that means that the turkey is absorbing 265 W of power.

Because the turkey is emitting 33 W but absorbing 265 W, the internal energy of the turkey is increasing and it is therefore warming. Eventually, the turkey reaches the temperature when it is considered "done," and the cook removes it from the oven. That is how a conventional oven cooks.

While the turkey is absorbing energy from the walls, by conservation of energy the walls must be losing energy and cooling down. The oven has a thermostat in it that senses this cooling and turns on the heating element to maintain the wall temperature at 375°F. This occasional cycling back on of the heating element is familiar to any cook.

A microwave oven also cooks food by bombarding food with photons. However, instead of bombarding the food with infrared photons, a microwave oven bombards the food with microwave photons, which have longer wavelengths. For reasons that we will not go into here, microwave ovens cook faster because they are able to deliver higher rates of power to the food than a conventional oven can. In the example here, the oven is delivering 265 W of power to the turkey. By using microwaves, however, the oven is able to deliver five to ten times that amount. The net result is that the food is heated more rapidly in a microwave oven than it is in a conventional oven.

I hope that you have a sense of the importance of the physics we have discussed in this chapter – it has a profound impact on your life and the world around you. I will show you in the next chapter that it also plays a key role in climate.

## 3.5 Chapter summary

- Energy is expressed in units of joules (J). Power is the rate that energy is flowing, and it is expressed in watts (W); 1 W = 1 J/s.
- Temperature is a measure of the internal energy of an object and is frequently expressed by physicists in units of Kelvin. The temperature in Kelvin is equal to the temperature in degrees Celsius plus 273.15.
- Photons are small discrete packets of energy. They have a characteristic size, known as the wavelength, which determines how the photons interact with matter. Photons with wavelengths between 0.3 and 0.8 m are visible to humans; photons with wavelengths between 0.8 and 1,000 μm are called infrared.
- Most objects emit blackbody radiation. The characteristic wavelength emitted by a blackbody is equal to $2,897/T$ (where wavelength is in microns and temperature is in Kelvin). The total power emitted per unit area by a blackbody is equal to $\sigma T^4$, where $\sigma = 5.67 \times 10^{-8}$ (W/m²)/K⁴ and temperature is in Kelvin. Photons

emitted by room-temperature objects are in the infrared and are not visible to humans.

- When a photon is emitted by an object and then absorbed by another object, this process transfers a small amount of energy from the emitter to the absorber.
- If the energy received by an object by absorbing photons exceeds the energy lost by emitting photons, then the object's internal energy increases – and it warms up. The object cools off if the energy in emitted photons exceeds the energy received by absorbing photons.

# Additional reading

D. Archer, *Global Warming: Understanding the Forecast*, 2nd ed. (Hoboken, NJ: Wiley, 2011). Chapter 2 of this excellent introductory text on climate change covers blackbody radiation.

For a more detailed discussion of blackbody radiation, see almost any introductory physics book.

See www.andrewdessler.com/chapter3 for additional resources for the chapter and www.andrewdessler.com/computer for computer exercises that illustrate some of the important concepts of radiation and energy balance.

# Terms

Blackbody
Electromagnetic radiation
Emissions spectra
Energy
Energy balance
Equilibrium
Incandescent light bulb
Infrared radiation
Internal energy
Joule
Kelvin scale
Micron
Photons
Power
Temperature
Ultraviolet
Visible photons
Watt
Wavelength
Wien's displacement law

# Problems

1. The temperature of an object goes up by 1 K. How much did it go up in degrees Fahrenheit and how much in degrees Celsius?
2. A sphere with a radius of 1 m has a temperature of 100°C. How much power is it radiating?
3. As a room-temperature object increases in temperature, it begins to glow. Describe the progression in colors as the object heats up. Ultimately, what happens to the glow if the warming continues to nearly infinite temperatures?
4. Consider two stars that have the spectra shown in Figure 3.7. Based just on the information provided in this plot, what are the colors and radiating temperatures of the stars? (The gray shading shows the range of wavelengths that humans can see.)
5. How much total energy (in watts) is the Sun radiating? It is a 6,000-K blackbody with a radius of 700,000 km.
6. You can dim an incandescent bulb by decreasing the temperature of the filament. What do you think happens to the color of the bulb as it dims? Find a dimmer and test your hypothesis.
7. If you run a 60-W light bulb for one week, how many joules of energy have been consumed?
8. Why are incandescent light bulbs being phased out in many countries (including the United States)?
9. The Sun as a blackbody:
   a. The Sun is a 6,000-K blackbody. At what characteristic wavelength does it radiate?
   b. How much power per unit surface area is the Sun radiating?
   c. Imagine that the Sun had a radius twice as large as it presently does, but emitted the same total amount of energy. What temperature would the Sun have to be?

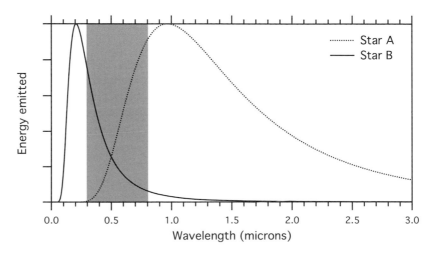

**Figure 3.7**    Emissions spectra of two hypothetical stars.

10. $E_{in}$ is the energy being absorbed by an object, and $E_{out}$ is the energy being radiated.

   a. If the temperature of an object is not changing, what does this tell us about $E_{in}$ and $E_{out}$?

   b. If the temperature of an object is increasing, what does this tell us about $E_{in}$ and $E_{out}$?

11. Your bank account has the same balance on April 1 as it did on March 1. Your friend suggests that this means that you did not deposit or withdraw any money for the entire month. Is that correct? Explain why or why not.

12. Heat capacity is the amount of energy it takes to warm up an object by 1 K. The heat capacity of water is 4.18 J/g/K; in other words, if you add 4.18 J to 1 g of water, the water will warm by 1 K. Imagine you have a cup containing 200 g of water that is absorbing 150 W of power.

   a. At what rate is the water warming? Answer is degrees K per second.

   b. If the cup starts at room temperature, how long would you have to heat it to reach boiling?

      To verify your answers, make sure the units work out.

13. I mentioned in the chapter how microwave ovens are able to deliver more energy to food during cooking than conventional ovens, so microwave ovens can cook food faster. For the following, imagine you are cooking a turkey in a conventional oven at 325 °F.

   a. What would you have to do in order to increase the amount of power being delivered to the turkey with the conventional oven?

   b. Would this cook the turkey faster? Why do we not cook turkeys that way?

   c. Why are microwave ovens able to deliver so much energy to food, while conventional ovens cannot?

      (to answer parts b and c, you have to know that the energy from a photon is absorbed by an object over a layer about one wavelength thick)

# 4

# A simple climate model

Scientists have been studying the Earth's climate for nearly 200 years and, over that time, a sophisticated and well-validated theory of our climate has emerged. In this chapter, we take the fundamental physics we learned in the last chapter and use it to explain how greenhouse gases warm the planet and why the temperature of the Earth is what it is. By the end of the chapter, you will understand why scientists have such high confidence that adding greenhouse gases to the atmosphere will warm the planet.

## 4.1 The source of energy for our climate system

The first step to understanding the climate is to do an energy budget calculation, which requires us to calculate the *energy in* and *energy out* for the Earth. The ultimate source of energy for our planet is the Sun, which puts out an amazing $3.8 \times 10^{26}$ W (380 trillion trillion W) of power. The Sun emits photons in all directions, so only a small fraction of this energy falls on the Earth. So the first step in calculating *energy in* is to determine the intensity of sunlight at the Earth's orbit.

To estimate this, imagine a sphere surrounding the Sun, with a radius equal to the Sun–Earth distance, 150 million km (Figure 4.1). Because the sphere completely encloses the Sun, all of the sunlight emitted by the Sun must fall on the interior of the sphere. The surface area of the sphere is $4\pi r^2 = 4\pi(150 \text{ million km})^2 = 2.8 \times 10^{17} \text{ km}^2 = 2.8 \times 10^{23} \text{ m}^2$. Dividing the total energy emitted by the Sun by the area of the sphere produces an estimate of the intensity of solar radiation at the Earth's orbit: $3.8 \times 10^{26} \text{ W}/2.8 \times 10^{23} \text{ m}^2 = 1{,}360 \text{ W/m}^2$. This value, $1{,}360 \text{ W/m}^2$, is known as the *solar constant* for the Earth; it is frequently represented in equations by the symbol $S$.

As should be obvious, the solar constant is a function of how far the planet is from the Sun. As a planet gets closer to the Sun, the solar constant for that planet increases; if it gets further away, the solar constant decreases. Our next-door neighbor Venus is located 107 million km from the Sun. Thus, the solar constant for Venus is $3.8 \times 10^{26}$ W divided by the surface area of a sphere with a radius of 107 million km, $1.43 \times 10^{23} \text{ m}^2$ – which yields a value of $2{,}600 \text{ W/m}^2$.

Now that we know the Earth's solar constant, we can determine the total solar energy falling on the Earth. The easiest way to quantitatively calculate this is to realize that, if we set up a screen behind the Earth, the Earth would cast a circular shadow on the screen, with a radius equal to the radius of the Earth (Figure 4.2). The amount of sunlight falling on the Earth is equal to the amount that would have fallen

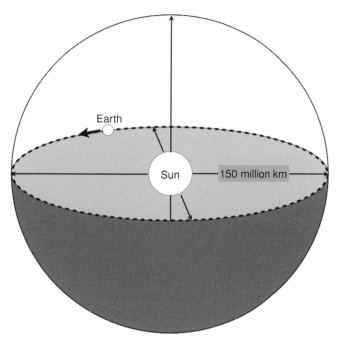

**Figure 4.1** Solar constant calculation: A sphere (gray) surrounds the Sun with a radius equal to the Earth's orbit (dashed line); all radiation emitted by the Sun (black arrows) falls on this sphere.

into the shadow area if the Earth were not there. The shadow area is $\pi R^2$, where $R$ is the radius of the Earth. So the total solar energy that would have fallen into that area is $\pi R^2$ times the solar constant $S$.

Given that the radius of the Earth is approximately 6,400 km = $6.4 \times 10^6$ m, and $S = 1,360$ W/m², solar energy is falling on the Earth at a rate of $1.8 \times 10^{17}$ W or 180,000 TW (1 TW, called a terawatt, is $10^{12}$ or a trillion watts). This is an immense amount of power. Human society today consumes about 16 TW, so this simple calculation shows why solar energy is the Holy Grail of renewable energy: If

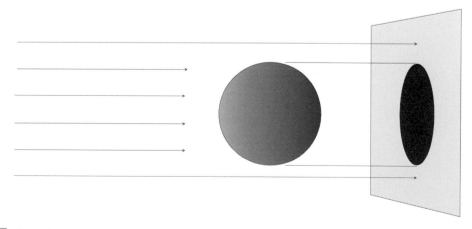

**Figure 4.2** The Earth is casting a shadow on a screen placed right behind it because it blocks sunlight. The total amount of solar energy falling on the Earth is the same as what would have fallen into the shadow area.

we could capture just 0.01 percent of the solar energy falling on the Earth, we could satisfy all of the world's current energy needs.

Not all of the photons from the Sun that fall on the Earth are absorbed by it. Some of the photons are reflected back to space by clouds, ice, and other reflective elements of the Earth system. The reflectivity of a planet is called the *albedo,* from the Latin word for "whiteness" (the word *albino* derives from the same root). It is frequently represented by the symbol $\alpha$, which is the fraction of incident photons that are reflected back to space; for the Earth, $\alpha$ is 0.3. This means that $1 - \alpha$ is the fraction of photons that are absorbed by the Earth. Taking this into account, *energy in* ($E_{in}$) for the Earth is

$$E_{in} = S(1 - \alpha)\pi R^2 \tag{4.1}$$

Evaluating Equation 4.1 yields an estimate of $E_{in}$ for the Earth of 120,000 TW. In the rest of the chapter, we will find it more useful to do the calculation per square meter of the Earth's surface area, so we divide Equation 4.1 by the surface area of the Earth, $4\pi R^2$:

$$\frac{E_{in}}{\text{area}} = \frac{S(1 - \alpha)\pi R^2}{4\pi R^2} = \frac{S(1 - \alpha)}{4} \tag{4.2}$$

Note that the $\pi R^2$ terms cancel, so the net amount of solar energy absorbed per square meter is *not* a function of the Earth's size. Plugging values of $S = 1360$ W/m$^2$ and $\alpha = 0.3$ into Equation 4.2, we obtain a value of 238 W/m$^2$ for the Earth's $E_{in}$. This is a good number to remember.

You might have noticed that I have become a bit sloppy with the terms *energy* and *power* in the previous discussion. In Equation 4.1, for example, the mathematical abbreviation for *energy in* appears on the left-hand side, but the right-hand side has units of power (Watts). The physics pedants will argue that we should be writing *power in* instead of *energy in,* and they are indeed correct. However, my choice of terminology here reflects the terminology actually used by scientists who do these kinds of energy-balance calculations. If you go to a meeting of climate scientists or read the peer-reviewed climate literature, you will find that they use *power* and *energy* interchangeably in equations like Equation 4.2. If you worry about things like this, then just keep track of the units and you will always know what is being talked about.

So the Earth absorbs an average of 238 W/m$^2$ from the Sun, but that does not mean that every square meter of the Earth absorbs this amount. In fact, the amount of solar energy absorbed varies widely across the planet. First, the nighttime half of the Earth receives no energy from the Sun at all. Second, the amount falling on a square meter of the daytime half is determined by the orientation of that square meter with respect to the incoming beams of sunlight. The amount of sunlight received is at maximum if the surface is oriented perpendicular to the incoming beam (Figure 4.3a). As the surface rotates away from perpendicular, the amount of solar energy intercepting the surface decreases (Figure 4.3b), eventually reaching zero for a surface parallel to the incoming beam (Figure 4.3c).[1]

---

[1] For those with a good grasp of geometry and geography, the energy falling on a square meter of surface varies as the cosine of the latitude.

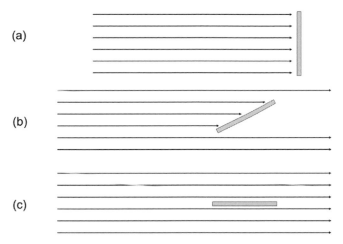

Figure 4.3 Schematic showing how the amount of energy falling on a surface is dependent on the angle between the surface and the incoming beams of light: (a) perpendicular, (b) rotated away from perpendicular, and (c) parallel.

Figure 4.4 shows how this leads to variations in the amount of solar energy falling on the Earth's surface with latitude. In the tropics (Arrow A), the surface of the Earth is perpendicular to the incoming solar light beams, corresponding to the situation in Figure 4.3a. The surface in the mid-latitudes (Arrow B) is at a moderate angle to the incoming solar light beams, corresponding to the situation in Figure 4.3b. This means that mid-latitudes receive less solar radiation per square meter than the tropics. Finally, the polar regions (Arrow C) correspond to the situation in Figure 4.3c, so this region receives even less solar energy.

In addition to variations in the incoming sunlight with latitude, the albedo of the planet also varies widely. The tropics are mainly open ocean, which is dark and therefore has a low albedo. Combined with the large amount of solar energy per square meter, the tropics therefore experience far more solar heating than anywhere else on the planet. The high latitudes, however, are generally covered by snow and ice, giving them a high albedo. Combined with the small amount of solar energy received per square meter, this means that the polar regions absorb the least amount of solar energy. This provides us with a simple but fundamentally correct explanation

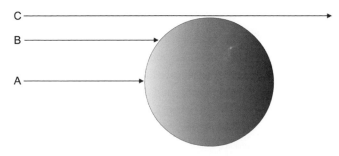

Figure 4.4 Schematic showing how the amount of solar energy falling on a square meter of the Earth's surface is determined by the latitude.

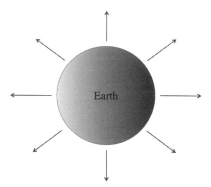

**Figure 4.5** Diagram of the Earth, showing the emission of infrared blackbody radiation in all directions.

of why the tropics tend to be the warmest place on the planet and the polar regions the coldest.

## 4.2 Energy loss to space

In the early nineteenth century, Joseph Fourier, one of history's great mathematicians, asked a deceptively simple question: Because energy is always falling on the Earth from the Sun, why does the Earth not heat up until it is the same temperature as the Sun? The answer he determined is that the Earth is losing energy at a rate equal to the rate at which it is receiving energy from the Sun.

On the basis of what we learned in Chapter 3, you may rightly guess that the Earth loses energy back to space by means of blackbody radiation (Figure 4.5). For a blackbody, $P/a$ is $\sigma T^4$, where $P/a$ is the power emitted per square meter, $T$ is the temperature of the planet, and $\sigma$ is the Stefan-Boltzmann constant, $5.67 \times 10^{-8}$ $(W/m^2)/K^4$. Setting $E_{in}$ (Equation 4.2) equal to $P/a$, the rate of *energy out*, we get the following equation:

$$\frac{S(1-\alpha)}{4} = \sigma T^4 \tag{4.3a}$$

Solving for $T$, we get

$$T = \sqrt[4]{\frac{S(1-\alpha)}{4\sigma}} \tag{4.3b}$$

Plugging $S = 1{,}360$ W/m$^2$ and $\alpha = 0.3$ into Equation 4.3b yields[2] a temperature $T = 255$ K ($-18\,^\circ$C). The actual average temperature of the Earth is closer to 288 K ($15\,^\circ$C), so our estimate of the Earth's temperature is too cold by $33\,^\circ$C. Where did our calculation go wrong? It turns out that what we have neglected is the heating of the

---

[2] The mathematical equation $a = \sqrt[4]{y}$ means that $a^4 = y$. To calculate the fourth root of $y$ on a calculator, you can use the $y^x$ key found on most calculators, where $x = 0.25$. A simpler way to calculate the fourth root of $y$ is to take the square root of $y$ and then take the square root of that number – in other words, $a = \sqrt{\sqrt{y}}$.

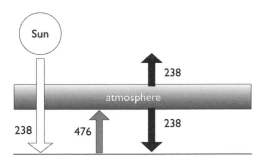

**Figure 4.6** Schematic of energy flow on a planet with a one-layer atmosphere. The atmosphere is represented by a single layer that is transparent to visible photons but absorbs all infrared photons that fall on it. The arrows show global average energy flows with values in W/m$^2$.

planet by the Earth's atmosphere, which is frequently referred to as the *greenhouse effect*.

## 4.3 The greenhouse effect

### 4.3.1 One-layer model

To understand the impact of the atmosphere on our planet's temperature, let us make the following assumptions (which turn out to be reasonably accurate).

1. The Earth's atmosphere is transparent to visible photons emitted by the Sun (which have wavelengths from 0.3–0.8 μm), so these photons pass through the atmosphere and are absorbed by the surface.
2. The atmosphere is opaque to infrared photons emitted by the surface (wavelengths longer than 4 μm), and so all of these photons are absorbed by the atmosphere.
3. The atmosphere also behaves like a blackbody, so it emits photons based on its temperature. It emits photons equally both upward and downward.
4. Photons emitted by the atmosphere in the upward direction escape to space and carry energy away from the Earth. Photons emitted downward are absorbed by the surface.

This one-layer model is diagramed in Figure 4.6. For conceptual simplicity, the diagram shows the effects of the atmosphere concentrated in a single thin layer, which is why this model is frequently called a "one-layer" model.

To calculate the surface temperature in this model, we assume that the planet as a whole, as well as the surface and the atmosphere individually, must all be in energy balance (where $E_{in}$ equals $E_{out}$). First, let us consider the energy balance for the planet as a whole. *Energy in* to the planet is coming entirely from the Sun. *Energy out* to space is coming entirely from the atmosphere (remember: any photons emitted by the surface are absorbed by the atmosphere; they do not escape to space). Using the Earth's values of solar constant and albedo, the *energy in* from the Sun is 238 W/m$^2$ (Equation 4.2). This means that the atmosphere must be radiating 238 W/m$^2$ upward to space in order for the planet as a whole to be in energy balance. Because the atmosphere radiates equally upward and downward, the atmosphere is also radiating 238 W/m$^2$ back toward the Earth's surface.

Now let us consider energy balance for the surface. *Energy in* for the surface is 238 W/m$^2$ from the Sun and 238 W/m$^2$ from the atmosphere, for a total of 476 W/m$^2$. This means that the surface has to be emitting 476 W/m$^2$ upward in order to achieve energy balance.

To make sure we did not make a mistake, we can check the energy balance for the atmosphere. *Energy in* comes from the surface, which is emitting 476 W/m$^2$. *Energy out* comes from emission of 238 W/m$^2$ upward to space and 238 W/m$^2$ downward to the surface, for a total *energy out* of 476 W/m$^2$. Thus, the atmosphere is indeed in energy balance.

So what is the temperature of the surface? If we know that the surface is emitting 476 W/m$^2$, we can determine its temperature by using the Stefan-Boltzmann equation from Chapter 3: $E_{out} = P/a = \sigma T^4$. Solving $\sigma T^4 = 476$W/m$^2$ for $T$ yields a surface temperature of 303 K (30 °C), which is 48 °C warmer than that for the planet without an atmosphere.

This is an incredibly important result: the addition of an atmosphere that is opaque to infrared radiation has significantly warmed the planet's surface. Conceptually, this occurs because the surface of the planet with an atmosphere is heated not just by the Sun but also by the atmosphere. Of course, if you walk outside, you cannot see the atmosphere heating the Earth's surface because the photons it emits are not visible, but they still carry energy. When scientists talk about the greenhouse effect, it is this heating of the surface by the atmosphere to which they are referring.

An alternative way to think about the greenhouse effect is that the atmosphere warms the surface by making it harder for the surface to lose energy to space. Without an atmosphere, all of the photons emitted by the surface escape to space; the surface has to emit only 238 W/m$^2$ for the planet to be in energy balance. With a one-layer atmosphere, though, only half of the photons emitted by the surface end up escaping to space – the other half are returned to the surface. This means that the surface must emit twice as much, 476 W/m$^2$, in order for 238 W/m$^2$ to escape to space. This higher rate of emission requires a warmer surface.

## 4.3.2 Two-layer model

Now let us consider a planet with a "two-layer" atmosphere (Figure 4.7). Once again, the atmosphere is transparent to visible radiation but opaque to infrared. That means that photons from the Sun pass through the atmosphere and are absorbed by the surface. Photons emitted by the surface are absorbed in the lower atmosphere. Photons emitted by the lower atmosphere in the upward direction are absorbed by the upper atmosphere; photons emitted in the downward direction are absorbed by the surface. Photons emitted by the upper atmosphere in the upward direction escape to space; photons emitted in the downward direction are absorbed by the lower atmosphere.

Once again, the key to determining the surface temperature is to enforce energy balance for the planet as a whole, the surface, and both atmospheric layers. The easiest way to do this is to start with planetary energy balance and then work downward from

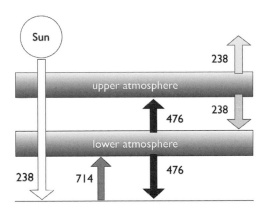

Figure 4.7  Schematic of energy flow on a planet with a two-layer atmosphere, with values in W/m².

the topmost layer to the surface. Planetary energy balance requires *energy out* for the planet to balance *energy in* from the Sun. Because *energy out* comes entirely from the upper layer, it must be emitting 238 W/m² to space in order to balance the 238 W/m² that the Sun is providing the planet (assuming terrestrial values for $S$ and $\alpha$). That, in turn, means that the upper layer is also emitting 238 W/m² downward.

Totaling the emissions in both directions, the upper layer is emitting 476 W/m². Because *energy out* must equal *energy in* for the layer, this layer must be receiving 476 W/m² from the lower atmospheric layer. Thus, we know that the lower layer is emitting 476 W/m² upward – and therefore downward, too. For the lower layer to achieve energy balance, the lower layer must also be receiving $476 + 476 = 952$ W/m². We already calculated that 238 W/m² are coming from the upper layer, so that means that 714 W/m² must be coming from the surface to the lower layer.

We can verify our result by examining the energy balance for the surface. The surface receives 476 W/m² from the lower atmosphere and 238 W/m² from the Sun, for a total $E_{in}$ of 714 W/m². This corroborates what we calculated must be $E_{out}$ for the surface based on energy balance for the lower atmosphere.

Finally, for the surface to be emitting 714 W/m², its temperature must be 335 K (62°C). This is 32°C warmer than the surface of the planet with a one-layer atmosphere and 80°C warmer than a planet with no atmosphere. Thus, adding a second layer to the atmosphere further increases the planet's surface temperature.

### 4.3.3 *n*-layer model

Now let us derive the surface temperature for a planet with $n$ layers (Figure 4.8). For some variety, let us assume that the planet has a solar constant $S = 2{,}000$ W/m² and an albedo $\alpha = 0.7$. Thus, *energy in* for this planet is $S(1 - \alpha)/4 = 150$ W/m². This means that upward emissions from the topmost layer of the atmosphere (Layer 1) must also be 150 W/m². And because upward and downward emissions must be the same, this layer is also emitting 150 W/m² downward, so that total *energy out* for this layer is 300 W/m². This in turn means that *energy in* for Layer 1 must also be 300 W/m².

*Energy in* for Layer 1 comes entirely from energy emitted by Layer 2. Layer 2 must therefore be emitting 300 W/m² upward. This means that it is also emitting

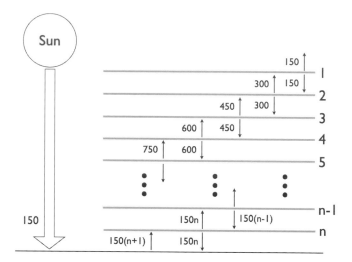

Figure 4.8 Schematic of energy flow on a planet with an n-layer atmosphere; layers are numbered from 1 to $n$ (topmost to bottommost layers), with values in W/m$^2$.

300 W/m$^2$ downward, for a total *energy out* of 600 W/m$^2$. *Energy in* for Layer 2 comes from downward emissions of Layer 1 and upward emissions of Layer 3 and must total 600 W/m$^2$ in order to balance *energy out*. Downward emissions from Layer 1 are 150 W/m$^2$, which means that upward emissions from Layer 3 must be 450 W/m$^2$.

Layer 3 must be emitting 450 W/m$^2$ both upward and downward, for a total *energy out* of 900 W/m$^2$. *Energy in* from downward emissions from Layer 2 is 300 W/m$^2$, meaning that upward emissions from Layer 4 must be 600 W/m$^2$.

By this time, a pattern has emerged and we can extrapolate to the bottommost layer, layer $n$. Layer $n$ is emitting $150\,n$ in both upward and downward directions. This in turn means that the surface is receiving $150\,n$ emitted from the bottommost layer and 150 W/m$^2$ from the Sun. For energy to balance, the surface must be emitting $150(n + 1)$ W/m$^2$ upward. Setting $150(n + 1) = \sigma T^4$, we can solve for the surface temperature $T$ of this planet:

$$T = \sqrt[4]{\frac{150(n + 1)}{\sigma}} \tag{4.4}$$

Figure 4.9 shows the surface temperature as a function of $n$, calculated by using Equation 4.4. As you increase the number of layers, the surface gets hotter and hotter. However, the warming is not linear – each additional layer produces less warming than the previous layer.

We can also write the general solution for the surface temperature of an $n$-layer planet:

$$T = \sqrt[4]{\frac{(n + 1)S(1 - \alpha)}{4\sigma}} \tag{4.5}$$

This is an important equation and one that is good to memorize. It says that the surface temperature of the planet is basically determined by three parameters: the number of layers in the atmosphere ($n$), the solar constant ($S$), and the albedo ($\alpha$).

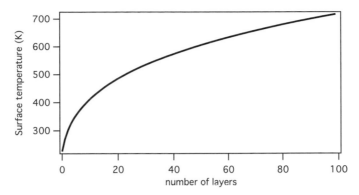

Figure 4.9 Surface temperature for the *n*-layer planet, as a function of the number of layers. Here, $S(1-\alpha)/4 = 150\,W/m^2$.

To connect this equation to the real world, I should make clear what the "number of layers" physically represents. As I will discuss in Chapter 5, it is the *greenhouse gases* in our atmosphere that absorb infrared photons. And the number of layers is equivalent to the amount of greenhouse gas in the atmosphere. Therefore, an increase in the amount of greenhouse gas in the atmosphere corresponds to an increase in the number of layers – and a warming climate.

Before we proceed to testing our theory, it is important to reiterate two important points:

1. We can now answer the question, "Why is the Earth's temperature what it is?" The temperature of our climate system is set by the requirement that *energy in* and *energy out* balance. And if $n$, $S$, or $\alpha$ change, then the temperature will adjust as required to reestablish this energy balance.
2. As you add more greenhouse gases to the planet (i.e., increase $n$), the temperature of the planet will increase.

## 4.4 Testing our theory with other planets

It is important to emphasize that the n-layer model discussed in Subsection 4.3.3 makes many simplifying assumptions. For example, not all energy transport on a planet is by radiation – some is transported by atmospheric motions, such as thunderstorms. Another simplification is that the model assumes an infinitely fast horizontal energy transport, allowing us to use a single temperature for the planet. In reality, though, transport of energy is slow enough that large temperature differences can develop between regions (e.g., the tropics and the polar regions or between night and day).

This means that you should not expect the model to produce quantitatively accurate surface temperatures. Nonetheless, the model captures the essential physics of our climate system. And, as I will show in this section, the model is successful in making

| Table 4.1 Data on the four inner planets in our solar system | | | | |
|---|---|---|---|---|
| Planet | Solar constant (W/m$^2$) | Albedo | Observed surface temperature (K) | Inferred $n$ |
| Mercury | 10,000 | 0.1 | 452 | 0.052 |
| Venus | 2,650 | 0.7 | 735 | 82 |
| Earth | 1,360 | 0.3 | 289 | 0.65 |
| Mars | 580 | 0.15 | 227 | 0.22 |

qualitative predictions of the relative surface temperatures of the Earth and its nearby neighbors, Mercury, Venus, and Mars.

Table 4.1 lists the important characteristics of the planets, and it reveals some puzzles. Mercury is the planet closest to the Sun, yet Venus, twice as far from the Sun as Mercury, has a surface temperature that is approximately 300 K warmer. This result becomes even more puzzling when we realize that, because of its high albedo, the *energy in* for Venus, $S(1 - \alpha)/4 = 200$ W/m$^2$, is more than a factor of ten smaller than that for Mercury (2,250 W/m$^2$). It is even less than the *energy in* for the Earth (238 W/m$^2$) – yet Venus is 450 K hotter than the Earth.

Given the surface temperature, albedo, and solar constant, we can solve Equation 4.5 for $n$, which is the number of layers required to satisfy energy balance. This "inferred $n$" is also listed in Table 4.1. The inferred $n$ for Mercury is near zero, suggesting it has almost no greenhouse effect. This is correct, because Mercury has essentially no atmosphere. For Mars, inferred $n = 0.22$. This again makes some sense – Mars has a thin atmosphere mostly containing carbon dioxide, so it does have some greenhouse effect. However, the Martian atmosphere has fewer greenhouse gases than the Earth's atmosphere, so the greenhouse effect on Mars is expected to be weaker than that on Earth. Our calculations confirm that.

Finally, our calculations suggest that Venus, with inferred $n = 82$, has a massive greenhouse effect. This is again correct. The surface pressure on Venus is ninety times that of Earth (1,300 psi, or pounds per square inch, compared with 14.5 psi here on Earth), and the atmosphere is mainly composed of carbon dioxide. The result of this massive, greenhouse-gas-rich atmosphere is a planet hotter than the inside of your oven on broil – hot enough even to melt lead. Thus, we see that Equation 4.5 successfully explains the relative climates of the innermost planets of our solar system.

## 4.5 Chapter summary

- In this chapter, we created a very simple climate model based on the fact that the solar energy received by a planet ($E_{in}$) must be balanced by the energy that is radiating to space ($E_{out}$). The temperature of the planet adjusts until this balance is achieved.
- For a planet, $E_{in} = S(1 - \alpha)/4$. $S$ is the solar constant, which is the intensity of sunlight at the planet's orbit (in units of W/m$^2$), and $\alpha$ is the planet's albedo,

which is the fraction of photons that fall on the planet that are reflected back to space.
- The *energy out* for a planet is due to blackbody radiation; $E_{out} = \sigma T^4$.
- In our simple model of the climate, the atmosphere is entirely transparent to visible radiation from the Sun, but it absorbs all infrared radiation. We can then calculate the surface temperature by enforcing energy balance for the surface, the atmosphere, and the planet as a whole.
- We derived a general equation, Equation 4.5, for the surface temperature $T$ of a planet. It is repeated here:

$$T = \sqrt[4]{\frac{(n+1)S(1-\alpha)}{4\sigma}}$$

- This equation says that the surface temperature of the planet is determined by three parameters: the number of layers in the atmosphere ($n$), which is a proxy for how much greenhouse gas is in the atmosphere, the solar constant ($S$), and the albedo ($\alpha$).
- This simple model also explains the relative temperatures of the Earth's nearest neighbors, namely Mercury, Venus, and Mars.

## Additional reading

For a more complete and technical description of the physics of climate, see the following books.

D. Archer, *Global Warming: Understanding the Forecast*, 2nd ed. (Hoboken, NJ: Wiley, 2011). Chapter 3 provides another description of the layer model at about the level of this textbook. Check it out to see how another author explains it.

J. T. Houghton, *The Physics of Atmospheres* (Cambridge: Cambridge University Press, 2001). This book covers climate physics at a level appropriate for an upper-level physics undergraduate.

R. T. Pierrehumbert, *Principles of Planetary Climate* (Cambridge: Cambridge University Press, 2011). This book is written at a level appropriate for a physics graduate student.

See www.andrewdessler.com/chapter4 for additional resources for the chapter and www.andrewdessler.com/computer for computer exercises that illustrate some of the important concepts of energy balance and layer models.

## Terms

Albedo
Greenhouse effect
Greenhouse gases
Solar constant

# Problems

1. What is the surface area of a sphere with radius $r$? What is the area of a disk with radius $r$? What is the area of a disk with diameter $d$?
2. A planet in another solar system has a solar constant $S = 2{,}000$ W/m$^2$, and the distance between the planet and the star is 100 million km.
   a) What is the total power output of the star? (Give your answer in watts.)
   b) What is the solar constant of a planet located 75 million km from the same star? (Give your answer in watts per square meter.)
3. Draw a diagram (like Figure 4.6) that shows the energy flows for a planet with a one-layer atmosphere. The solar constant for the planet is $S = 900$ W/m$^2$, and the albedo of the planet is $\alpha = 0.25$. Make sure each arrow is labeled with the energy flow. What is the surface temperature of this planet?
4. Draw a diagram (like Figure 4.7) that shows the energy flows for a planet with a two-layer atmosphere. The solar constant for the planet is $S = 3{,}000$ W/m$^2$ and the albedo of the planet is $\alpha = 0.1$. Make sure each arrow is labeled with the energy flow. What is the surface temperature of this planet?
5. Two people argue about why Venus is so much warmer than the Earth. The first argues that it is because Venus is closer to the Sun, so it absorbs more solar energy. The second argues that it is because Venus has a thick, greenhouse-gas-rich atmosphere. Which person is right, and why is the other one wrong?
6. Some recently discovered planets in other solar systems are so hot that they glow in the visible; they are literally "red hot" (e.g., do a Google search for "HD 149026b").
   a) How many atmospheric layers would the Earth need before it glowed in the visible? (Assume $S = 1{,}360$ W/m$^2$ and $\alpha = 0.3$.) To answer this, you must first estimate what temperature the Earth has to be to begin glowing.
   b) Alternatively, what would the solar constant have to increase to for a one-layer planet with an albedo $\alpha = 0.3$?
   c) How far would the Earth have to be from the Sun in order to have this solar constant?
7. Assume a planet with a one-layer atmosphere has a solar constant $S = 2{,}000$ W/m$^2$ and an albedo $\alpha = 0.4$.
   a) What is the planet's surface temperature? Make the standard assumption that the atmosphere is transparent to visible photons but opaque to infrared photons.
   b) During a war on this planet, a large number of nuclear weapons are exploded, which kicks enormous amounts of dust and smoke into the atmosphere. The net result is that the atmosphere now absorbs visible radiation – so solar energy is now absorbed in the atmosphere. It also still absorbs infrared radiation. Draw a diagram like Figure 4.6 to show the fluxes for this new situation, and calculate the planet's surface temperature. The solar constant and albedo remain unchanged.
   c) Explain in words why the temperature changes the way it does after the nuclear war. Is describing this as "nuclear winter" appropriate?

8. Assume a planet with a one-layer atmosphere and values of solar constant $S = 1{,}000$ W/m$^2$ and albedo $\alpha = 0.25$. Let us assume there is some dust in the atmosphere, so that 50 percent of the Sun's energy is absorbed by the atmosphere and 50 percent by the surface. Draw a diagram like Figure 4.6 to show the fluxes, and calculate the planet's surface temperature.

9. Derive an expression for the fraction of energy received by the surface that comes from the atmosphere (this is the amount of energy that comes from the atmosphere divided by the sum of energy from the Sun and energy from the atmosphere). Using values in Table 4.1, calculate the fraction for Mercury, Earth, and Venus. Make the standard assumption that the atmosphere is transparent to visible photons but opaque to infrared photons.

10. On Mercury, which has no atmosphere, the difference in temperature between daytime and nighttime temperatures can be 700 K. On the Earth, the difference between daytime and nighttime temperatures can be 30 K. On Venus, there is basically no difference between daytime and nighttime temperatures. Why is this? (If you get stuck, working Question 9 might help you answer this question.)

11. As we will discover in Chapter 11, one way to solve global warming is to increase the reflectivity of the planet (I will explain how later). To reduce the Earth's temperature by 1 K, how much would we have to change the albedo? (assume a one-layer planet with an initial albedo of 0.3 and solar constant of 1360 W/m$^2$).

12. Given fixed $n$ and $\alpha$, how does the temperature of a planet vary with r, the distance between the planet and the star? Hint: Work out how $S$ varies with $r$, and plug that into Equation 4.5.

13. A planet has a solar constant $S = 2{,}000$ W/m$^2$, an albedo $\alpha = 0.7$, and a radius $r = 3{,}000$ km. What would happen to the temperature if the planet's radius doubles?

14. One argument you hear against mainstream climate science is that adding greenhouse gases to the atmosphere is like painting a window. Eventually, the window is opaque, so that adding another coat of paint does nothing. Is this a good analogy? Is there a point where adding greenhouse gases to the atmosphere does not lead to increases in the planet's temperature?

15. Imagine that the Sun's radius doubles (but the Sun maintains the same surface temperature). What happens to the Earth's solar constant and surface temperature?

16. Explain how the variation of solar energy with surface orientation (Figure 4.3) can explain the variation in local temperature through the day. The warmest temperatures during the day are usually found from 3 PM to 5 PM; does this fit with your theory?

17. If you were on a spacecraft and you pointed an infrared thermometer at a one-layer planet, what temperature would it read? What if the planet had two layers? Or $n$-layers? Assume $S$ and $\alpha$ are the same as for the Earth. Remember that an infrared thermometer measures the radiation coming from an object and yields the temperature the object must be in order to be emitting that radiation.

18. Newly formed stars are often obscured by the dense dust clouds from which they form. To see how these appear to observers on the Earth, imagine that fifty stars, each identical to our Sun, form in a spherical cloud of dust with a radius of 100 billion km. Much like the atmosphere, the dust absorbs all of the light given off by the stars and radiates an equal amount of energy to the rest of the Universe. What temperature does the cloud appear to be from outside the cloud? What wavelength telescope would you need to see the dust cloud?

# 5

# The carbon cycle

In the simple model of the climate presented in Chapter 4, the temperature of a planet is set by the number of atmospheric "layers," the albedo, and the solar constant. I said there that the number of layers is determined by the abundance of greenhouse gases in the atmosphere, but I was intentionally vague about what a greenhouse gas is, or which components of our atmosphere are greenhouse gases. In this chapter, I address these questions and discuss in detail one of our atmosphere's most important greenhouse gases, carbon dioxide.

Carbon dioxide, or $CO_2$, is the primary greenhouse gas emitted by human activities, and policies to control modern climate change frequently focus on reducing our emissions of this gas. But constructing rational climate change policies requires more than just knowing how much of it humans are dumping into the atmosphere. It requires an understanding of the *carbon cycle* – how carbon moves between the atmosphere, ocean, land biosphere, and rocks on the Earth. This will help us understand what happens to carbon dioxide after it is emitted into the atmosphere, which in turn will help us understand the future trajectory of our climate.

## 5.1 Greenhouse gases and our atmosphere's composition

As we learned in Chapter 4, the greenhouse effect occurs because our atmosphere is mostly transparent to visible photons but absorbs infrared photons. It turns out that only a few of the components of our atmosphere actually absorb infrared photons, and it is these *greenhouse gases* that are responsible for the Earth's greenhouse effect. In this section, I describe the composition of our atmosphere, with a particular focus on greenhouse gases.

Approximately 78 percent of the dry atmosphere ("dry atmosphere" excludes water vapor)[1] is made up of diatomic nitrogen or $N_2$ – two nitrogen atoms bound together. About 21 percent is diatomic oxygen or $O_2$, which is two oxygen atoms bound together; this is the part of the atmosphere that we need to breathe to survive. Argon atoms make up approximately 1 percent of our atmosphere. None of these three constituents, which together make up more than 99 percent of the dry atmosphere, absorbs infrared photons, so they are not greenhouse gases; therefore, they do not warm the surface of the planet.

---

[1] Throughout this section, the percentages given are of volume, not mass. For the chemists reading this, this is the same as mole fraction.

The next biggest component of the atmosphere is water vapor or $H_2O$, a constituent whose abundance varies widely from place to place. In the warm tropics, water vapor can make up as much as 4 percent of the atmosphere. In cold polar regions, in contrast, water vapor may be only 0.2 percent. Its abundance decreases rapidly with altitude, and in the stratosphere it typically makes up 0.0005 percent of the atmosphere.

Water vapor is the most abundant and important greenhouse gas in our atmosphere. Its main source is evaporation from the oceans, and it is primarily removed from the atmosphere when water forms raindrops and these fall to the surface. Emissions of water vapor from human activities contribute essentially nothing to its atmospheric abundance. In Chapter 6, I will talk in more detail about the role water vapor plays in climate change and how humans are indirectly increasing its abundance.

Taken together, diatomic nitrogen and oxygen, water vapor, and argon make up more than 99.95 percent of the atmosphere. You might expect the remaining 0.05 percent to have no important role because it seems like such a small amount, but you would be wrong. This last smidgen of atmosphere is crucial to life on the planet.

The largest part of this remaining 0.05 percent is carbon dioxide or $CO_2$, which made up 0.04 percent of the atmosphere in 2014. Carbon dioxide absorbs infrared photons and is therefore a greenhouse gas. In fact, it is the second most important greenhouse gas, behind water vapor. Because 0.04 percent is an awkwardly small number, scientists typically express the concentration of these trace gases in a more convenient unit: *parts per million.* A concentration expressed in parts per million indicates how many molecules out of every million are the gas in question.[2] In this case, 0.04 percent corresponds to 400 parts per million or ppm, meaning that there are 400 molecules of carbon dioxide in every million molecules of air.

Parts per million can be usefully contrasted with percent, which indicates how many molecules of every 100 are the species in question. In fact, the word *percent* comes from the marriage of the words *per cent,* literally meaning "out of 100." Thus, air is approximately 78 percent nitrogen, which means that 78 out of every 100 molecules of air are molecules of nitrogen.

The next most important greenhouse gas in our atmosphere is methane or $CH_4$. In 2014, it had an atmospheric abundance of 1.83 ppm. Despite its small abundance, methane is also a key player in our climate; I will discuss it in detail in Section 5.7.

Another important greenhouse gas in our atmosphere is nitrous oxide or $N_2O$, which is present in today's atmosphere at concentrations of about 0.32 ppm. This molecule is also known as "laughing gas," which your dentist might give you before she works on your teeth. It is emitted into the atmosphere from nitrogen-based fertilizer and industrial processes as well as several natural sources.

*Ozone* is another greenhouse gas. Its chemical formula is $O_3$, so it is a molecule made up of three oxygen atoms. The abundance of ozone varies widely across the atmosphere – in unpolluted air near the surface, its abundance is about 10-40 parts per billion,[3] whereas its abundance can reach 10 ppm in the stratosphere – 1,000 times higher.

---

[2] Parts per million can be by volume (number of molecules out of every million) or by mass (grams of constituent out of every million grams of air). Following the previous discussion, all mixing ratios in this book will be by volume.

[3] This means that, of every billion molecules of air, a few are ozone.

Ozone is absolutely necessary for life on our planet because it absorbs high-energy ultraviolet photons emitted by the Sun before they reach the Earth's surface. These photons carry enough energy that they can seriously damage living tissue – leading to diseases such as skin cancer in humans. But ozone is also one of the primary components of photochemical smog, and breathing it can lead to health problems in humans and animals; ground-level ozone can also damage plants. Thus, you want ozone between yourself and the Sun, but you do not want to breathe it. Because of this, ozone high up in the stratosphere is considered "good" ozone, whereas ozone near the ground is "bad" ozone.

A final group of greenhouse gases are the *halocarbons*, including chlorofluorocarbons and hydrochlorofluorocarbons, which are synthetic industrial chemicals used as refrigerants (e.g., in air conditioners and refrigerators) and in various industrial applications. This category also includes natural molecules such as methyl chloride. Together, they are present in today's atmosphere at a concentration of a few parts per billion, and all of them are powerful greenhouse gases. These halocarbons are also the main culprits behind ozone depletion.

These greenhouse gases are not equal in their ability to warm the planet. Methane is roughly twenty times more powerful than carbon dioxide on a per molecule basis – meaning that it takes twenty molecules or so of carbon dioxide to equal the warming from one molecule of methane. The most powerful greenhouse gases on a per molecule basis are the halocarbons. It takes several thousand carbon dioxide molecules to equal the warming from one halocarbon molecule – so despite being about 1/10,000th as abundant as carbon dioxide, these halocarbons nevertheless make an important contribution to the greenhouse effect. Do not forget, too, that the three most abundant gases in our atmosphere, namely diatomic nitrogen, diatomic oxygen, and argon, which together make up about 99.9 percent of the dry atmosphere, are not greenhouse gases at all. These differences in warming potential among various gases have important implications when designing policies to address climate change.

As you will see, carbon dioxide is the most important greenhouse gas for the problem of modern climate change. Because of this, this chapter will mainly focus on it and the processes that regulate its atmospheric abundance, which are collectively known as the *carbon cycle*.

## 5.2 Atmosphere-land biosphere-ocean carbon exchange

### 5.2.1 Atmosphere-land biosphere exchange

We have been directly monitoring the abundance of carbon dioxide in the atmosphere since the middle of the twentieth century. Figure 5.1 plots two years (twenty-four months) of measurements, showing that the amount of carbon dioxide in the atmosphere varies throughout the year: during its maximum in May, carbon dioxide is about 6 ppm higher than the September minimum.

This annual cycle in carbon dioxide reflects the annual cycle of plant growth and decay. Plants absorb carbon dioxide from the atmosphere and use it to produce more

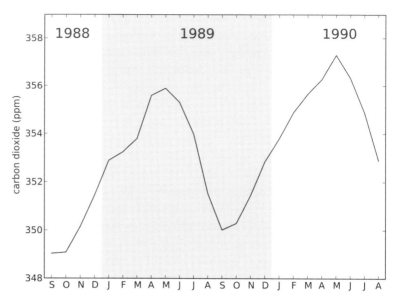

**Figure 5.1** The atmospheric abundance of carbon dioxide from fall 1988 through fall 1990 (data measured at Mauna Loa, Hawaii, and obtained from the NOAA Earth System Research Laboratory/Global Monitoring Division; see ftp://aftp.cmdl.noaa.gov/products/trends/co2/co2_mm_mlo.txt).

plant material in a process known as *photosynthesis:*

$$CO_2 + H_2O + \text{sunlight} \rightarrow CH_2O + O_2 \tag{5.1}$$

In this reaction, carbon dioxide, water, and sunlight combine to produce $CH_2O$ and $O_2$. Because $CH_2O$ is a combination of carbon and water, molecules made up of this unit are generally referred to as carbohydrates – in this context, you can think of $CH_2O$ as the chemical formula for a plant. Diatomic oxygen produced in this reaction is released into the atmosphere. This is the main source for the oxygen in our atmosphere, which we breathe to survive.

At the same time, humans, animals, and bacteria consume plant material in order to produce energy through a reaction known as *respiration:*

$$CH_2O + O_2 \rightarrow CO_2 + H_2O + \text{energy} \tag{5.2}$$

The net result of Equation 5.2 is carbon dioxide, which is released back into the atmosphere, and energy, which is used to power the organism. It should be noted that Equations 5.1 and 5.2 do not represent actual chemical reactions; rather, they represent the net of a large number of complex biochemical reactions that occur within the cells of organisms.

Equation 5.2 is the reverse of Equation 5.1: the carbon dioxide consumed in the production of the plant material in Equation 5.1 is released back into the atmosphere when the plant is consumed in Equation 5.2. Similarly, the oxygen molecule produced in Equation 5.1 is consumed in Equation 5.2. The production of a carbohydrate through photosynthesis followed by its consumption during respiration therefore produces no net change in either carbon dioxide or oxygen. Instead, the net effect is the conversion of sunlight into energy to power living creatures.

The atmosphere contained approximately 850 gigatonnes of carbon (GtC) in 2014. A gigatonne is 1 billion metric tons, where 1 metric ton is 1,000 kg or 2,200 lbs. Note that this is just the mass of the carbon atoms in the atmosphere – although the carbon dioxide molecule also contains two oxygen atoms, their mass is not included. Unfortunately, you will also see the mass expressed as the mass of carbon dioxide, which does include the mass of the two oxygen atoms. In that case, the atmosphere contains more than 3,100 $GtCO_2$ (billion metric tons of carbon dioxide). You can convert between these units by using the fact that 1 GtC $= 3.67$ $GtCO_2$. In this book, I will use GtC exclusively, but you must be careful to identify whether the mass is given in GtC or $GtCO_2$ when you read anything about climate change.

The land biosphere contains 2,500 GtC, stored in living plants and animals and in organic carbon in soils (e.g., decaying leaves). During a given year, photosynthesis removes approximately 120 GtC from the atmosphere. Respiration roughly balances this, transferring about the same amount back to the atmosphere. Thus, over a year, there are only small changes in carbon dioxide in the atmosphere or land biosphere as a result of photosynthesis or respiration.

The fact that photosynthesis and respiration are balanced over the year does not mean that they are in balance at every point in time. Most of the Earth's land area – and, therefore, most of the Earth's plants – are found in the northern hemisphere. During the northern hemisphere's spring and summer (May-September), when plants are growing and trees are leafing, global photosynthesis exceeds respiration and there is a net drawdown of carbon dioxide out of the atmosphere and into the land biosphere; we can see this in Figure 5.1.

During the northern hemisphere's fall and winter (October-April), plant material that was produced during the spring and summer decays, releasing carbon dioxide back into the atmosphere. During this period, global respiration exceeds photosynthesis and there is a net transfer of carbon from the biosphere into the atmosphere, which we can also see in Figure 5.1.

There is also a large amount of carbon stored in permafrost, which is ground that is frozen year-round. Much like that frozen dinner that has been in your freezer since Bill Clinton was President, dead organic plant matter frozen into the permafrost does not decay; it is kept intact as long as the ground remains frozen. If permafrost thaws out, however, the organic matter stored there will begin to decay, releasing carbon into the atmosphere.

While permafrost is indeed melting, it appears at present that this contributes little to atmospheric carbon dioxide. However, given that much of this permafrost is in the Arctic, which is expected to continue warming rapidly, the melting will continue. At some point, perhaps soon, the resulting release of carbon dioxide may contribute significantly to atmospheric carbon dioxide.

## An aside: Where does the oxygen in our atmosphere come from?

As I mentioned earlier, photosynthesis followed by respiration is not a net producer or consumer of carbon dioxide or oxygen. Where, then, does the large amount of molecular oxygen in our atmosphere come from? It turns out that it is the result of photosynthesis that is not balanced by respiration. That occurs when a plant grows through photosynthesis, but the plant material is buried before it can be consumed via respiration. When that happens, the oxygen produced during photosynthesis is not consumed. Over the billions of years that life has existed on the planet, this process has built up and now maintains the oxygen levels in our atmosphere.

## 5.2.2 Atmosphere-ocean carbon exchange

One of carbon dioxide's most important properties is that it readily dissolves in water. Once it has dissolved in water, carbon dioxide is converted to *carbonic acid* ($H_2CO_3$) by means of this reaction:

$$CO_2 + H_2O \rightarrow H_2CO_3 \tag{5.3}$$

This process is sometimes referred to as ocean acidification. As the oceans become more acidic, the biology of the oceans can change – and given human reliance on the oceans for food, this could lead to important impacts on humans. This will be discussed in more detail in Chapter 9.

The carbonic acid formed in Equation 5.3 can react further with water to convert into other forms of carbon. Because of the conversion of carbon dioxide to many other forms of carbon, the ocean can absorb huge amounts of carbon dioxide. Carbon is returned to the atmosphere in a reaction that is the reverse of Equation 5.3:

$$H_2CO_3 \rightarrow CO_2 + H_2O \tag{5.4}$$

This is followed by the escape of carbon dioxide back into the atmosphere. Processes embodied by Equations 5.2 and 5.3 transfer about 80 GtC per year between the atmosphere and ocean, roughly similar to the exchange between atmosphere and land biosphere.

Thus, carbon cycles easily between the atmosphere and ocean. To fully understand this exchange, however, we must think of the ocean as being split into two parts. The first part is the top 100 m or so of the ocean, which exchanges carbon very rapidly with the atmosphere. This part of the ocean makes up only a few percent of the mass of the ocean and is often referred to as the *mixed layer* because it is well mixed by winds and other weather events; it contains 900 GtC.

Below this lies the other 97 percent of the ocean, known as the *deep ocean*. The deep ocean also contains most of the ocean's carbon, approximately 40,000 GtC, or forty-seven times more carbon than is in the atmosphere. The mixed layer and the deep ocean exchange carbon at a rate of about 100 GtC per year. This occurs as ocean currents mix high-carbon water from the mixed layer with low-carbon water from the deep ocean. It also occurs when sinking organic matter, such as dead organisms or fecal material, falls from the mixed layer into the deep ocean – a process known as the biological carbon pump.

### 5.2.3  The combined atmosphere-land biosphere-ocean system

Figure 5.2 shows a schematic of the combined atmosphere-land biosphere-ocean system. Approximately 120 GtC per year are continuously cycling between the

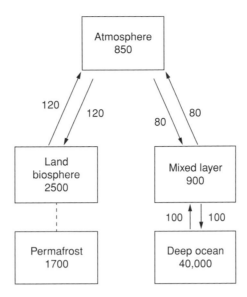

Figure 5.2 A schematic of exchange between the atmosphere, land biosphere, and ocean. Reservoirs are in GtC; fluxes are in GtC per year. Based on Figure 6.1 of Ciais et al. [2013].

atmosphere and land biosphere as plants absorb carbon dioxide as they grow and then release carbon dioxide when they die. About 80 GtC per year of carbon from the atmosphere is continuously dissolving into the ocean's mixed layer, while the same mass of carbon atoms is coming out of the ocean and back into the atmosphere, thereby cycling between the atmosphere and ocean. At the same time, the mixed-layer and deep ocean are exchanging 100 GtC per year.

To get an idea of what these numbers actually mean, we calculate *turnover times* for the atmosphere and land biosphere reservoirs. The turnover time for the atmosphere is the length of time that a carbon atom in the atmosphere will remain there before being transferred into one of the other two reservoirs. This can be roughly estimated as the size of the reservoir, 850 GtC, divided by the total flux out of the reservoir, 200 GtC per year (120 GtC per year goes into the land biosphere and 80 GtC per year goes into the mixed layer). This yields an atmospheric turnover time of about four years. This turnover time is also referred to as a "lifetime" or "residence time."

This means that a carbon atom stays in the atmosphere for only four years or so before it is transferred into the land biosphere or ocean. Remember that this is an average value – an individual molecule of carbon dioxide may remain in the atmosphere for a shorter or longer time. Another way to think about a turnover time is that, over a period of four years, enough exchange will take place to replace all of the carbon that is in the atmosphere with carbon from the land biosphere or ocean.

The turnover time of carbon in the land biosphere is 2,500 GtC divided by 120 GtC per year = 21 years. This means that a carbon atom in the land biosphere will stay there for 21 years before being transferred into the atmosphere. Thus, it takes a few decades for a carbon atom to make a round trip from the land biosphere into the atmosphere and back into the land biosphere.

We can also calculate the turnover times for the ocean reservoirs. The total flux out of the mixed layer is 180 GtC per year (80 GtC per year is exchanged with the

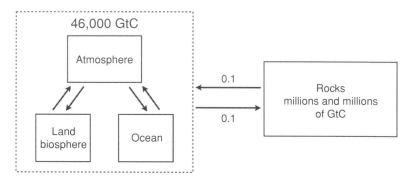

Figure 5.3 A schematic of exchange between the atmosphere-land biosphere-ocean reservoir and the rock reservoir (reservoirs are in GtC; fluxes are in GtC per year).

atmosphere and 100 GtC per year is exchanged with the deep ocean), so the turnover time for the mixed layer is 900 GtC ÷ 180 GtC per year ≈ five years. The turnover time for the deep ocean is several centuries: 40,000 GtC ÷ 100 GtCr ≈ 400 years. Thus, it takes a few centuries for a carbon atom to make a round trip from the atmosphere through the mixed layer, the deep ocean, and back.

Another way to think about this is that the atmosphere exchanges carbon rapidly (*time scale* of years to decades) with the land biosphere and mixed layer, and much more slowly (time scale of centuries) with the deep ocean. Later in the chapter, I will explain why this is so important for the climate change problem.

## 5.3 Atmosphere-rock exchange

Most of the carbon in the world – many millions of gigatons of carbon – is stored in rocks, such as limestone ($CaCO_3$), and this carbon is slowly exchanging with the atmosphere-land biosphere-ocean system (Figure 5.3). Carbon dioxide is transferred from rocks directly into the atmosphere by volcanic eruptions. This process releases an average of 0.1 GtC per year. Although this flux is small compared with other fluxes, over millions of years it can lead to significant transfers of carbon into the atmosphere-land biosphere-ocean system.

These natural emissions of carbon dioxide from the rock reservoir are roughly balanced by a process known as *chemical weathering*, which removes about an equal amount of carbon from the atmosphere and transfers it back into rocks. Chemical weathering starts when carbon dioxide in the atmosphere dissolves into raindrops falling toward the surface (remember that carbon dioxide dissolves readily in water). Carbonic acid ($H_2CO_3$) is produced via Equation 5.3, which makes the rain slightly acidic (pH = 5.6).

When this acidic rain falls on rocks, both the physical impact of the rain and chemical reactions break the rock down. The chemical reaction is shown here:

$$CaSiO_3 + CO_2 \rightarrow CaCO_3 + SiO_2 \tag{5.5}$$

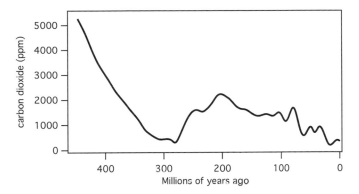

**Figure 5.4** Atmospheric carbon dioxide over the past half-billion years (based on Royer, 2006, Figure 1).

Note that this equation is a general description of the process of weathering, not the exact chemical reaction. Nonetheless, the essential message of Equation 5.5 is correct: The carbon dioxide molecule consumed in this reaction came from the atmosphere, via rainwater, and it is transferred into a molecule of calcium carbonate or $CaCO_3$, which is limestone, and subsequently runs off with the rainwater and eventually reaches the ocean. The reaction also forms silicon dioxide ($SiO_2$), the primary component of sand, quartz, and glass.

Once in the ocean, the molecules of calcium carbonate are deposited through various mechanisms on the sea floor. Over many millions of years, plate tectonics carries this calcium carbonate deep within the Earth, where high temperatures and pressures turn the rock into magma. Eventually, this carbon is transferred back to the surface by volcanism, thereby releasing the carbon dioxide back to the atmosphere and completing the cycle.

Another pathway for carbon to move into the rock reservoir occurs when plants are rapidly buried in sediment before they can decay. This is, in fact, the same process that leads to a net production of oxygen (discussed earlier). Once buried and subjected to the great heat and pressure found deep within the Earth, this dead plant material can be converted to fossil fuels, which humans extract and burn for energy – thereby returning the carbon to the atmosphere. We will explore fossil fuels in the next section.

A carbon atom will remain in the atmosphere-land biosphere-ocean system for approximately 46,000 GtC ÷ (0.1 GtC per year) = 460,000 years before it is transferred into the rock reservoir. Given the large size of the rock reservoir and the relatively small rate of exchange between the rocks and the atmosphere, it takes many, many millions of years for a carbon atom to travel through the rock reservoir and reemerge into the atmosphere.

Figure 5.4 shows an estimate of atmospheric carbon dioxide over the past half-billion years. Four hundred million years ago, atmospheric carbon dioxide was more than ten times higher than it is today, and since that time it has generally decreased, although there have been wide variations. The abundance in 2014, 400 ppm, is relatively low when compared to the geologic record (although it is higher than it has been for several million years).

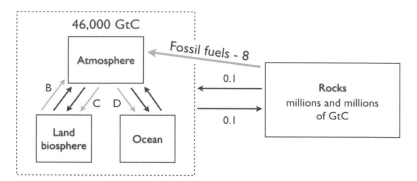

**Figure 5.5** Diagram of the carbon cycle as perturbed by humans. Gray arrows show net flows of carbon caused by human activities. Arrows B, C, and D represent deforestation, enhanced absorption of carbon by the land biosphere, and enhanced absorption of carbon by the ocean, respectively.

The variations in atmospheric carbon dioxide in Figure 5.4 are due to variations in the rate of exchange of atmospheric carbon with the rock reservoir. This includes variations in the rate at which carbon dioxide is emitted from volcanoes – during periods of extreme volcanism, for example, atmospheric carbon dioxide will increase.

The movement of the continents is another factor. As continents move, patterns of rainfall can change, and new rock can be exposed to the atmosphere, both of which can change the rate of chemical weathering – and therefore the rate at which carbon dioxide is removed from the atmosphere. For example, approximately 40 million years ago, the Indian subcontinent collided with the Asian continent, forming the Himalayas and the adjacent Tibetan Plateau. Changing wind patterns brought heavy rainfall onto the expanse of newly exposed rock, and the resultant chemical weathering has been drawing down atmospheric carbon dioxide since then – until humans came along, that is.

## 5.4 How are humans perturbing the carbon cycle?

As Figure 5.4 shows, carbon dioxide varies without any human activities. However, humans can also affect the carbon cycle. Figure 5.5 shows the perturbed carbon cycle, with the flows of carbon caused by human activities indicated as the gray lines. The main perturbation comes from the combustion of *fossil fuels* for energy. Fossil fuels were formed when plants that grew hundreds of millions of years ago were buried before the carbon in them could be released back into the atmosphere by respiration. Under high pressure and heat, applied over millions of years, the carbon in the plants was converted into the substances we know today as oil, coal, and natural gas.

When fossil fuels are burned, the net reaction is similar to the respiration reaction (Equation 5.2):

$$CH_x + O_2 \rightarrow CO_2 + H_2O + energy \tag{5.6}$$

Fossil fuels are represented in Equation 5.6 by $CH_x$ because they are primarily carbon, with varying amounts of hydrogen. During combustion, the fossil fuel combines with oxygen to produce energy, carbon dioxide, and an amount of water vapor that depends on how much hydrogen was in the fuel. The resulting energy is used to power our world, and the carbon dioxide is vented directly into the atmosphere. Fossil fuels can also contain other trace species, such as sulfur. When burned, these trace species can also be released into the environment and lead to environmental problems of their own, such as acid rain. Also note that Equation 5.6 is a schematic reaction, not an actual chemical reaction, so do not be concerned that it does not balance.

Before humans discovered them, fossil fuels were safely sequestered in the rock reservoir. The natural carbon cycle would have slowly released this carbon back to the atmosphere through geologic processes over many millions of years. Humans, however, are extracting and combusting fossil fuels at a breath-taking pace – fast enough that we will extract and burn most of the world's fossil fuels in just a few hundred years. The net result is the creation of an additional pathway for carbon from rocks to the atmosphere (the line marked "Fossil fuels" in Figure 5.5). And this pathway is large: over the period 2002–2011, the combustion of fossil fuels led to average emissions of approximately 8.3 GtC per year to the atmosphere.[4] This is more than eighty times the natural flow rate of carbon from the rock reservoir to the atmosphere.

Humans have also been chopping down large tracts of forest – a process known as *deforestation* – in order to use the land for other activities, such as agriculture or grazing livestock. Frequently, the forest is removed by burning it, which releases the carbon stored in trees and other plants to the atmosphere. Even just bulldozing the forest releases the carbon to the atmosphere, albeit more slowly. In addition, the soil often contains large amounts of carbon stored in the form of dead organic plant material. When the forest is removed, much of this plant material decomposes following Equation 5.2, releasing the carbon back into the atmosphere.

Deforestation is just one of many ways that man's impact on the land can influence atmospheric carbon dioxide. Emissions associated with these changes are known collectively as *land-use changes*. Land-use changes are an important source of carbon dioxide for the atmosphere, and estimates are that it contributed approximately 0.9 GtC per year to the atmosphere from 2002 to 2011 – about one-tenth of the emissions from fossil fuel combustion. This flux is shown in Figure 5.5 as Arrow B.

Figure 5.6a shows that the abundance of carbon dioxide in our atmosphere has remained in a narrow range, 260–280 ppm, over the last 10,000 years – until a sudden spike occurred in the past few hundred years. Figure 5.6b focuses on this spike, showing that atmospheric carbon dioxide began rapidly increasing around 1800. This coincides with the industrial revolution, when widespread burning of fossil fuels began.

Figure 5.6b also shows that the rise in carbon dioxide is accelerating. Over the past 250 years, atmospheric carbon dioxide has increased by about 120 ppm. The first 60 ppm of the increase took more than 200 years, until 1980, whereas the next 60 ppm took only thirty-five years. Figure 5.6c shows high-resolution measurements of the abundance of carbon dioxide in our atmosphere over the past fifty-five years. The yearly sawtooth pattern reflects the seasonal cycle in plant growth, which was discussed in Section 5.2.1. There is also a long-term increase in carbon dioxide, from

---

[4] Cement production is lumped into this number; it constitutes a few percent of this total.

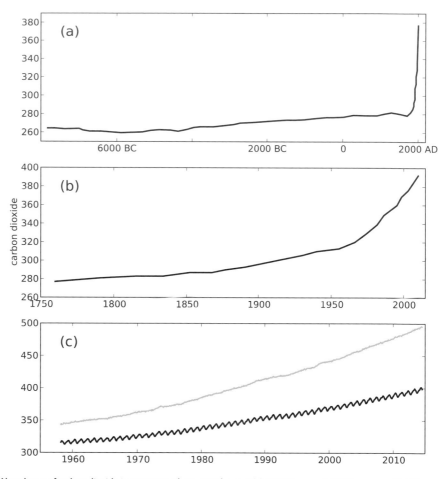

**Figure 5.6** Abundance of carbon dioxide in our atmosphere over the past (a) 10,000 years, (b) 250 years, and (c) 50 years (sawtooth line). Panel c also shows (gray curve) the annual average carbon dioxide abundance if all of the carbon dioxide emitted by human activities since 1959 had remained in the atmosphere. Panels a and b are adapted from Figure SPM.1 of IPCC, 2007a; measurements in panel c are obtained from the NOAA Earth System Research Laboratory/Global Monitoring Division, at http://www.esrl.noaa.gov/gmd/ccgg/trends/; the 100 percent-retained line is calculated by assuming that 55 percent of the emitted carbon is rapidly removed.

315 ppm in the late 1950s to 400 ppm in 2014, caused by fossil fuel combustion and land-use changes.

Figure 5.7 shows the year-to-year increase in atmospheric carbon dioxide. In the late 1950s, the increase in atmospheric carbon dioxide was less than 1 ppm per year. By the first decade of the twenty-first century, atmospheric carbon dioxide was increasing by nearly 2 ppm per year. This reflects the increasing emissions due to increasing fossil fuel combustion over the past half-century.

We are emitting carbon dioxide to the atmosphere, and we can see that the amount of carbon dioxide in the atmosphere is increasing. So far, so good. But there is a problem. We have good records of exactly how much fossil fuel is extracted and burned each year, so we can calculate how much atmospheric carbon dioxide *should*

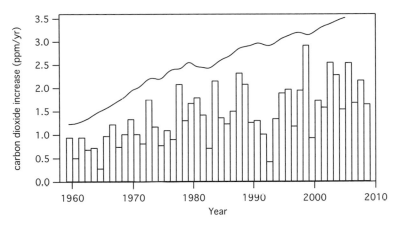

**Figure 5.7**    Bars show the observed year-to-year increase in atmospheric carbon dioxide. The solid line shows what the annual increase would have been had 100 percent of the carbon dioxide emissions remained in the atmosphere (bars are provided by the NOAA Earth System Research Laboratory/Global Monitoring Division, at http://www.esrl.noaa.gov/gmd/ccgg/trends/; the solid line is from Denman et al., 2007, Figure 7.4).

have increased. During 2002–2011, human emissions of carbon to the atmosphere averaged 9.2 GtC per year. But during that time, the increase in atmospheric carbon dioxide averaged 4.3 GtC per year – thus, the increase in carbon dioxide was about half of what we were emitting.

Figure 5.7 shows this more clearly: the solid curve is an estimate of how much atmospheric carbon dioxide should have increased if it had all stayed in the atmosphere. Over the past half century, the increase in atmospheric carbon dioxide each year is approximately half of the amount emitted.

So where are the rest of our emissions going? It turns out that about half of this "missing carbon" is dissolving into the ocean. This is indicated as Arrow D in Figure 5.5. The other half is believed to have gone into the land biosphere, although there is considerable scientific debate about exactly what part of the land biosphere is absorbing the carbon. This enhanced land sink is represented by Arrow C in Figure 5.5.

An aside: How do we measure how much carbon is going into the ocean versus the land-biosphere?

The curve of observed atmospheric carbon dioxide abundances (Figure 5.6c) is often referred to as the *Keeling Curve*, named after Charles D. Keeling, the scientist who initiated the measurements in 1957. The acorn, as they say, never falls far from the tree, and Keeling's son, Ralph, is following in his father's footsteps by making long-term measurements of another key atmospheric measurements: diatomic oxygen, $O_2$.

Figure 5.8 shows Ralph Keeling's measurements of oxygen over the last twenty-five years, along with measurements of carbon dioxide. The annual sawtooth, due to the seasonal cycle of plant growth, is apparent in both time series. However, the plots are exactly out of phase – for example, when the carbon dioxide is going up, oxygen is going down. Equations 5.1 and 5.2 explain why this is so: when carbon dioxide is absorbed by a plant during photosynthesis, oxygen is released to the atmosphere;

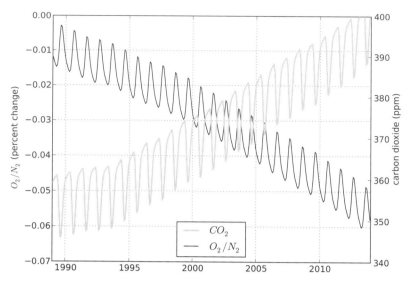

**Figure 5.8**  Time series of the change (in percent) of the $O_2/N_2$ ratio (relative to a standard) and $CO_2$ mixing ratio (in ppm) at the Canadian Alert Station (82°N). Diatomic nitrogen ($N_2$) is not changing much, so most of the change in the ratio is due to changes in oxygen. The main cause of the overall downward trend in $O_2/N_2$ is the loss of $O_2$ due to fossil fuel burning, which is also the main cause of the overall rise in $CO_2$.

when carbon dioxide is released during respiration, oxygen is absorbed from the atmosphere.

Keeling's data also show a long-term decline of oxygen in the atmosphere. This decline is the result of burning fossil fuels (see Equation 5.6); given the mix of fuels we use, about 1.4 molecules of oxygen are consumed for every carbon dioxide molecule released. So the good news is that we finally found a gas in the atmosphere whose abundance is declining – the bad news is that it is oxygen![5]

The exact amount of oxygen lost gives us important information about how much carbon the land and ocean are absorbing because the different pathways lead to different changes in oxygen. To see how this works, imagine that some fossil fuel is burned, and this adds 100 molecules of carbon dioxide to our atmosphere; this would also remove 140 molecules of oxygen.

As discussed earlier, about half of the carbon dioxide (in this case, fifty molecules) is absorbed by some combination of the land biosphere and ocean. Absorption by the ocean does not affect oxygen (Equation 5.3), so combustion followed by ocean absorption would lead to a net decrease in oxygen of 140 molecules.

Absorption by the land biosphere, however, produces oxygen (Equation 5.1) – about 1.1 molecules of oxygen are produced for every carbon dioxide molecule that is absorbed during photosynthesis. If all fifty carbon dioxide molecules are absorbed by the land biosphere, then $50 \times 1.1 = 55$ molecules of oxygen would be produced. So the net change in oxygen in this case would be a net loss of $140 - 55 = 95$ molecules of oxygen.

---

[5]  By the way, there is so much oxygen in the atmosphere that this small observed decrease should not worry you.

From Keeling's oxygen measurements, we find that the decline in oxygen falls about midway between these limits, leading to the conclusion that about half of the carbon is going into the land biosphere and half into the ocean.

Figure 5.7 also shows large variability in the year-to-year increase of atmospheric carbon dioxide. For example, atmospheric carbon dioxide increased by nearly 3 ppm in 1998, but less than 1 ppm the very next year – even though emissions were roughly the same. This is mainly due to variations in the climate due to El Niño events. Regional climate variations during these events modify the uptake and emissions of carbon from the land biosphere by varying areas of rainfall and drought.

If the land biosphere and ocean were not taking up about half of the carbon we emit, then atmospheric carbon dioxide would be much higher today than it actually is. Figure 5.6c also shows a rough estimate of what the long-term time series of atmospheric carbon dioxide would be if all emissions had remained in the atmosphere – it shows that atmospheric carbon dioxide might be near 500 ppm in 2014, approximately 100 ppm higher than the actual abundance. The climate would consequently be warmer and changing even more rapidly. Thus, the land biosphere and ocean are doing us a huge favor by absorbing significant amounts of carbon emitted by humans.

An emerging concern is whether the oceans and land biosphere can continue taking up as much carbon in the future as they presently are. It is unknown when or if we will reach a saturation point at which point the reservoirs slow down or even cease their uptake. If that happens, then a higher fraction of emissions will remain in the atmosphere, and the abundance of carbon dioxide in our atmosphere will grow more rapidly. This leads to yet another worry for climate scientists: that climate change itself may alter the carbon cycle. In the next chapter, we will explore how such carbon cycle feedbacks may amplify climate change.

## 5.5 Some commonly asked questions about the carbon cycle

Because of its central role in the climate change problem, climate skeptics occasionally challenge the claim that the observed increase in carbon dioxide since 1800 is due to human activities. In this section I address this argument.

> How do we know that combustion of fossil fuels is responsible for the increase in carbon dioxide, rather than nonhuman sources such as volcanoes or plants?

This is a reasonable question. After all, the amount of carbon dioxide absorbed by plants during the year and balanced by plant decay (approximately 120 GtC per year) is much larger than human emissions (which averaged 9 GtC per year over 2002–2011) – ditto for the ocean fluxes. So it may seem reasonable that the increase in atmospheric carbon dioxide might be driven by a slight excess of plant respiration over photosynthesis, or a slight excess flux of carbon dioxide out of the ocean. Similarly, we know that volcanoes emit carbon dioxide, and that over millions of years, volcanoes are a primary source of carbon dioxide in the atmosphere. So maybe the increase in atmospheric carbon dioxide is due to enhanced volcanic activity.

There are, however, several independent lines of evidence that unanimously agree that fossil fuel combustion is the dominant reason for the increase in atmospheric carbon dioxide over the past few centuries. First, Figure 5.7 shows that, for the past half-century, each year's increase in carbon dioxide in the atmosphere has been on average about half of what humans released into the atmosphere in that same year. Thus, when humans were emitting smaller amounts of carbon dioxide in the 1960s, atmospheric carbon dioxide was increasing at a slower rate than when humans were dumping large amounts of carbon dioxide in the atmosphere, as we are today. If the source of carbon dioxide emissions were nonhuman, it seems unlikely that it would track human emissions of carbon dioxide so closely.

Second, the carbon dioxide can be chemically "fingerprinted" to show that it comes from fossil fuels. The method is based on *isotopes* of carbon. All carbon atoms have six protons, but carbon's isotopes have different numbers of neutrons. The most abundant isotope is carbon-12, containing six neutrons to go with the six protons, and which makes up roughly 99 percent of the carbon on Earth. Carbon-13, with seven neutrons, makes up 1 percent of the carbon, and approximately one carbon atom out of a trillion is carbon-14, which has eight neutrons.

The chemical properties of an atom are for the most part set by the number of protons, so isotopes tend to have very similar chemical properties. The chemistry, though, is not identical. Plants, for example, preferentially absorb carbon-12 when growing. And because fossil fuels are derived from plants, they reflect this preference for carbon-12. When the fossil fuels are burned, the carbon dioxide produced also reflects plants' preference for carbon-12 over carbon-13.

Scientists can measure the amount of carbon-12 and carbon-13 in atmospheric carbon dioxide, and those measurements show that the increase in atmospheric carbon dioxide in Figure 5.6b is caused by carbon that is depleted in carbon-13 – such as that which comes from plants. This allows us to rule out sources such as volcanoes or the ocean. The measurements of atmospheric oxygen discussed earlier are also inconsistent with a volcanic or oceanic source.

Thus, we know that the increase in atmospheric carbon dioxide is coming from plants, but is it coming from plants that died hundreds of millions of years ago (i.e., fossil fuels) or plants of today? In order to make that determination, we turn to carbon-14. Carbon-14 is produced in the atmosphere when a neutron created by a cosmic ray hits the nucleus of an atom of nitrogen-14. The nucleus absorbs the neutron and ejects a proton, thereby transforming itself into carbon-14. Carbon-14 atoms are incorporated into molecules of carbon dioxide and are then absorbed by plants and incorporated into plant material. If you walk outside and pull a leaf off a tree, a small fraction of atoms in that leaf would be carbon-14.

Carbon-14 is known as radiocarbon because it is radioactive. That means its nucleus is unstable and converts back to nitrogen-14 with a half-life of approximately 6,000 years (so that, after 6,000 years, half of the carbon-14 has converted back to nitrogen-14). To see the implications of this, imagine a cotton plant that grew 6,000 years ago. As it grew, the plant absorbed carbon dioxide containing carbon-14 from the atmosphere, and the carbon-14 was incorporated into the plant. Immediately after the plant was picked, it would have the same proportion of carbon-14 as any living plant, and so would the cotton produced from it. But because it was no longer alive, it stopped absorbing carbon dioxide from the atmosphere. Over time, the amount of carbon-14 in the cotton slowly decreased as it was converted back to nitrogen-14.

Now imagine that modern-day archaeologists find a blanket made of this cotton and want to know how old it is. To do this, they measure the proportion of carbon-14 in the blanket and find that it has half the carbon-14 of a living plant. With a half-life of 6,000 years, the archaeologists conclude that the blanket is 6,000 years old. If they found that it had one fourth of the carbon-14 of a living plant, then it would be 12,000 years old. This process is known as *radiocarbon dating*.

Now let us turn our attention to fossil fuels. As we learned earlier, fossil fuels are produced when plant matter is buried for millions of years. After millions of years of being underground, all of the carbon-14 has converted back to nitrogen-14. Thus, fossil fuels contain essentially no carbon-14, a condition known as *radiocarbon dead*. So when the fossil fuels are burned, the carbon dioxide produced also has no carbon-14 in it. Scientists measuring the isotopic composition of atmospheric carbon dioxide have found that the carbon dioxide being added to the atmosphere is indeed radiocarbon dead, showing that it is coming from long-dead plants – fossil fuels – and not modern plants.

Putting all of the evidence together, along with an absence of any counterevidence, we see that there is no question that human activities are increasing the amount of carbon dioxide in the atmosphere. As my colleague John Nielsen-Gammon puts it, not only can we see the smoking gun, but the smoke is a chemical match to the gunpowder.

> Why focus on carbon dioxide from fossil fuel combustion when plants and animals emit far more carbon dioxide to the atmosphere?

Humans, animals, bacteria, and plants do indeed emit enormous amounts of carbon dioxide to the atmosphere – the land biosphere emits 120 GtC per year, compared with present-day emissions from human activities of about 9 GtC per year. So why should we care about carbon dioxide from fossil fuels? To understand the answer, you need to understand the difference between carbon dioxide coming from fossil fuel combustion and from respiration by living organisms.

Let us begin by imagining that you plant a carrot seed, and over the next few months this seed grows into a carrot. As described by Equation 5.1, the plant grows by absorbing carbon dioxide directly from the atmosphere, and this reduces the amount of carbon dioxide in the atmosphere. Now let us imagine that the carrot is eaten by a goat. The goat metabolizes the carrot (in a manner approximately following Equation 5.2), which produces energy to power the goat's vital functions. The carbon dioxide produced is exhaled back into the atmosphere.

Thus, when an animal exhales carbon dioxide, it is releasing back into the atmosphere carbon dioxide that was in the atmosphere just a few months before. Although this can lead to seasonal variations in carbon dioxide, as shown in Figure 5.1, it does not cause long-term increases in carbon dioxide. Figure 5.6a confirms this by showing basically no change in carbon dioxide over the past 10,000 years – a period during which humans, plants, and animals were certainly releasing carbon dioxide to the atmosphere.

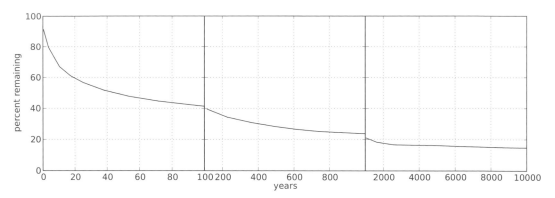

Figure 5.9 Fraction of carbon dioxide remaining in the atmosphere after an initial pulse in year zero. The plot shows that it takes a very long time for carbon dioxide emitted to the atmosphere to be completely removed. Based on Figure 1 of Box 6.1 of Ciais et al. (2013).

In contrast, when you burn fossil fuels, you are releasing to the atmosphere the carbon dioxide that had been safely sequestered in rocks (e.g., Figure 5.5) for hundreds of millions of years. This is a net addition to the atmosphere, so it does cause a long-term increase in carbon dioxide. Figure 5.6b confirms this: The increase in atmospheric carbon dioxide started with the industrial revolution, when society-wide burning of fossil fuels began.

## 5.6 The long-term fate of carbon dioxide

To get a feel for the long-term evolution of the climate over the next millennium, we need to know how long the carbon dioxide we release stays in the atmosphere. As a thought experiment, imagine that a pulse of carbon dioxide is instantaneously released into the atmosphere. At first, the land biosphere and mixed-layer of the ocean rapidly take carbon dioxide out of the atmosphere – Figure 5.9 shows that about 40 percent of the carbon dioxide pulse is removed in twenty years. Removing additional carbon dioxide requires transport into the deep ocean, which is a slower process. After 400 years, about 25 percent of the slug of carbon dioxide pulse remains in the atmosphere.

At this point, the deep ocean is in equilibrium with the atmosphere and cannot absorb any more carbon. Further removal of carbon dioxide requires reactions between carbon dissolved in the ocean and calcium carbonate ($CaCO_3$) sea floor sediments. These reactions transfer carbon to the ocean sediments, allowing the ocean to absorb more carbon dioxide. But this process is very slow – after 10,000 years, 15 percent of the initial pulse of carbon dioxide is still in the atmosphere. The last 15 percent is removed by chemical weathering (Equation 5.5) over the next few hundred thousand years.

The very long time it takes for carbon to be removed from the atmosphere is confirmed by estimates of carbon dioxide during the Paleocene-Eocene Thermal

Maximum (discussed in Section 2.2.2; see Figure 2.11). This is an event about 55 million years ago when a huge pulse of carbon (several thousand GtC) was released into the atmosphere, leading to a sudden and significant warming of the planet. It took several hundred thousand years for that slug of carbon to be removed and for the warming it caused to dissipate.

The upshot is that, if we add carbon dioxide to the atmosphere, it will remain in the atmosphere for a very long time, and the warming it causes therefore also sticks around for a very long time. We will revisit the grim implications of this in Chapter 8, but it is useful to realize that the actions we take in the next few decades will determine the trajectory of the climate for thousands, if not tens of thousands of years. If our actions this century lead to massive, long-term climate change, I wonder what people living in much warmer climates in the years 3000, 4000, or 10,000 will think of us.

# 5.7 Methane

Most discussions of the carbon cycle focus on the cycling of atmospheric carbon dioxide. However, methane is another crucial carbon-containing gas. Although the atmospheric abundance of methane was only 1.83 ppm in 2014, much smaller than that of carbon dioxide, on a per molecule basis, methane is roughly twenty times more powerful a greenhouse gas than carbon dioxide.

Methane is emitted to the atmosphere from both human and natural processes. About 60 percent of human methane emissions are from agriculture and waste. Livestock is the largest source of methane in this category. Cattle, as well as goats and sheep, are ruminants, and these animals produce methane in their guts during the digestion of food. This methane is eventually released to the atmosphere (out of both ends of the animals). The next largest source in this category is bacterial processes in landfills and other waste repositories. Emissions from rice paddies are the third significant source. In the warm and wet environment of a flooded rice field, bacteria in the soil efficiently produce methane, the vast majority of which is then released to the atmosphere.

The release of methane from the petrochemical industry is responsible for about 30 percent of human emissions of methane. This comes from leakage of methane from natural gas wells as well as release of geologic methane from coalmines. Finally, burning of forest and other biomass primarily produces carbon dioxide, but it also produces methane if the combustion temperature is sufficiently low (e.g., a smoldering fire). This is responsible for the remaining 10 percent of human methane emissions.

During the 2000s, methane emissions from natural sources were about equal to human emissions. Approximately two-thirds of these natural emissions were from natural wetlands, which produce methane the same way that flooded rice paddies do. Minor contributions come from the ocean, from freshwater lakes and rivers, and from wild animals, particularly termites.

Figure 5.10 shows that the atmospheric abundances of methane began rising about 1800, the same point at which carbon dioxide began rising. Scientists have also

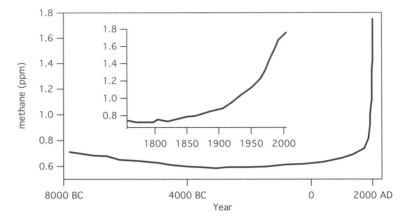

**Figure 5.10** Methane abundance over the past 10,000 years. The inset plot shows a close-up of the past 250 years (adapted from the IPCC, 2007a, Figure SPM.1).

attributed this increase to human activities. As we will see in Chapter 6, these emissions are enough to make methane a significant contributor to the warming we are experiencing – methane's contribution to global warming is approximately one-fourth of the contribution of carbon dioxide. As a result, reductions of methane emissions are frequently included in plans to address climate change.

Methane is removed from the atmosphere by oxidation, which follows the following schematic reaction:

$$CH_4 + 2O_2 \rightarrow CO_2 + 2H_2O$$

On average, a molecule of methane is destroyed by this reaction ten years after it was emitted. If we stopped emitting methane today, within a few decades, all of the human-emitted methane would be gone, and the atmospheric abundance would be back down to preindustrial amounts. This is quite different from carbon dioxide, which can stay in the atmosphere for centuries or millennia.

# 5.8 Chapter summary

- Only a few components of our atmosphere are greenhouse gases, which absorb infrared photons. The three most important are (in order) water vapor, carbon dioxide, and methane. Nitrogen, oxygen, and argon, which make up approximately 99.9 percent of the dry atmosphere, are not greenhouse gases.
- The carbon cycle describes how carbon cycles through its primary reservoirs: the atmosphere (containing 850 GtC), land biosphere (2,500 GtC), ocean (900 GtC in the mixed layer and 40,000 GtC in the deep ocean), and rocks (millions and millions of GtC).
- The atmosphere exchanges carbon with the land biosphere through photosynthesis and respiration. The atmosphere exchanges carbon with the ocean when carbon dioxide dissolves into or is emitted from the ocean. Once in the ocean, the carbon

dioxide is converted to carbonic acid and other chemicals. Over the course of several centuries, a carbon atom added to the atmosphere will cycle through all of the other reservoirs and return to the atmosphere.

- The atmosphere-land biosphere-ocean system also exchanges carbon with rock reservoirs through volcanism and chemical weathering. This exchange is extremely slow.
- Humans are perturbing the carbon cycle by extracting and burning fossil fuels. The result is the creation of a new, rapid pathway moving carbon from rocks to the atmosphere. Between 2002 and 2011, fossil fuel combustion released an average of 8.3 GtC to the atmosphere from the rock reservoir, which is more than eighty times the amount released from the rocks naturally. Land-use change is another important human source, releasing 0.9 GtC per year from the land-biosphere into the atmosphere during this period.
- This has increased atmospheric carbon dioxide abundance from approximately 280 ppm in 1750, before the industrial revolution, to 400 ppm in 2014.
- It takes a long time for the carbon cycle to remove carbon that humans add to the atmosphere. About 40 percent of it is removed in a few decades, 75 percent in a few centuries, and the last 25 percent is removed over tens and hundreds of thousands of years. This means that atmospheric carbon dioxide will be elevated by human activities for a very long time – even if we stop burning fossil fuels in the next few decades.
- Methane is another important greenhouse gas – each molecule of methane has the warming power of approximately twenty carbon dioxide molecules. In the last decade, about half of the methane emissions are due to human activities. These human emissions have increased atmospheric methane from approximately 0.8 ppm in 1750, before the industrial revolution, to 1.83 ppm in 2014. Methane's lifetime in the atmosphere is ten years, which is much shorter than carbon dioxide.

## Additional reading

Ciais et al., "Carbon and Other Biogeochemical Cycles," in T. F. Stocker, D. Qin, G.-K. Plattner, M. Tignor, S. K. Allen, J. Boschung, A. Nauels, Y. Xia, V. Bex, and P. M. Midgley (eds.), *Climate Change 2013: The Physical Science Basis*. Contribution of Working Group I to the Fifth Assessment Report of the Intergovernmental Panel on Climate Change (Cambridge and New York: Cambridge University Press, 2013). This is the carbon cycle chapter from the latest IPCC report. As with most IPCC reports, it is not the easiest thing to read, but if you want to know what scientists think, this is where you should go.

D. Archer, *The Global Carbon Cycle* (Princeton, NJ: Princeton University Press, 2010). This is a short and focused textbook on carbon cycle science. It is a great introduction to the subject.

E. Roston, *The Carbon Age: How Life's Core Element Has Become Civilization's Greatest Threat* (New York: Walker, 2009). This is a fun and easy-to-read book about carbon and the immense role it plays in our lives.

See www.andrewdessler.com/chapter5 for additional resources for this chapter.

# Terms

Carbon cycle
Carbonic acid
Chemical weathering
Deep ocean
Deforestation
Fossil fuels
Greenhouse gas
Halocarbons
Isotopes
Keeling curve
Land-use changes
Mixed layer
Ozone
Parts per million
Photosynthesis
Radiocarbon dating
Radiocarbon dead
Respiration
Time scale
Turnover time

# Problems

1. a) Describe the processes that transfer carbon from the atmosphere to the land and from the land to the atmosphere. What are the chemical reactions that describe these processes?
   b) How do these processes interact to produce the "sawtooth" annual cycle in the atmospheric abundance of $CO_2$ shown in Figures 5.1 and 5.6?

2. A letter to the editor of the *Austin American-Statesman,* published on December 23, 2009, asks this question: "The trillion-dollar question that Copenhagen has not answered [is this]: Because carbon dioxide molecules are all identical, why is it that carbon dioxide from carbonated beverages, pets, cattle, farm animals, and humans, yeast, dry ice, fireplaces, charcoal grills, campfires, wildfires, alcohol and ethanol is good, and carbon dioxide from fossil fuel is bad? Can anyone in the United States answer this question?" What is your answer?

3. Your aunt asks you how we know that volcanoes are not responsible for the observed increase in carbon dioxide. What do you tell her?

4. Explain how isotopes help us identify human activities as the reason atmospheric carbon dioxide is increasing.

5. Your grandfather asks you to explain how humans are modifying the carbon cycle. What do you tell him?

6. Explain how "chemical weathering" removes $CO_2$ from the atmosphere. What is the weathering chemical reaction? Can this process play an important role in counteracting the increase in atmospheric carbon dioxide caused by humans?

7. Of the carbon dioxide humans add to the climate, approximately half is removed within a few decades. Where does it go? How would it affect the climate if, all of the sudden, all of the carbon dioxide we emit stayed in the atmosphere?

8. Why is rain naturally acidic? What then, does the term *acid rain* refer to? (Acid rain is not covered in the chapter, so you will have to do some outside research on it.)

9. Imagine that 100 molecules of carbon dioxide are produced by combustion of fossil fuels; half of these molecules are immediately absorbed by the land biosphere and ocean. A fraction $f$ of this goes into the land biosphere (so $1-f$ goes into the ocean). a) Make a plot of the change in oxygen as $f$ varies from 0 to 1. b) If the observed change in oxygen is a decrease of 110 molecules, what fraction of carbon was absorbed by the ocean? Assume that fossil fuel combustion consumes 1.4 molecules of $O_2$ for every molecule of $CO_2$ produces; and photosynthesis produces 1.1 molecules of $O_2$ for every molecule of $CO_2$ consumed; absorption of $CO_2$ by the ocean has no effect on $O_2$.

10. The sawtooth in the $CO_2$ time series due to the annual cycle in northern hemisphere plant growth is dramatic. The data in Figure 5.1 comes from Mauna Loa, in Hawaii. Based on the material in this chapter, predict how the magnitude of this annual cycle in the Arctic would compare to that in the Antarctic. Find the data online (http://andrewdessler.com/chapter5) and see if you can confirm your hypothesis.

11. Why do the atmospheric oxygen measurements disprove an oceanic or volcanic source for the increase in atmospheric carbon dioxide?

# 6

# Forcing, feedbacks, and climate sensitivity

In Chapter 4, we showed that the temperature of a planet is a function of the solar constant, the albedo of the planet, and the composition of the atmosphere (Equation 4.5). In Chapter 5, we showed that humans are adding greenhouse gases to the atmosphere, so we would expect the planet's temperature to be increasing. In Chapter 2, we showed that temperature is indeed going up. If that were all there was to climate change, we would be done with the science. But there is a lot more interesting physics that we have to consider to fully understand modern climate change.

## 6.1 Time lags in the climate system

In our climate calculations in Chapter 4, we discussed equilibrium situations in which we explicitly assume that $E_{in}$ and $E_{out}$ are equal. But modern climate change is not an equilibrium problem. To understand the time-dependent behavior of the climate system, consider a planet with no atmosphere that is in equilibrium (*energy in* equals *energy out*). Assuming values for the Earth ($E_{in} = 238$ W/m$^2$), we calculated in Chapter 4 that the planet's surface temperature would be 255 K. The energy fluxes for this planet are diagramed in Figure 6.1a.

Now let us imagine that a one-layer atmosphere is instantly added to the planet. What are the fluxes the instant after the layer is added? The temperature of the surface is still 255 K because objects have *thermal inertia*, which prevents their temperatures from changing instantly – just like the inertia that keeps a car from stopping instantly when you hit the brakes. Anyone who has worked in a kitchen knows that turkeys placed in an oven do not cook instantly but take hours to heat up.

This means that, the instant after the atmosphere is added, the surface is emitting exactly the same as it was before the atmosphere was added, 238 W/m$^2$. The atmosphere, however, is now absorbing all of the photons coming from the surface. Half of the absorbed energy is reemitted upward to space, and half is reemitted downward back to the surface.[1] This is shown in Figure 6.1b.

The addition of the atmosphere has therefore reduced *energy out* for the planet by half, to 119 W/m$^2$. Given that *energy in* remains the same, 238 W/m$^2$, *energy in* exceeds *energy out* for the planet, so the planet must warm in response. Moreover,

---

[1] I am implicitly assuming here that *energy in* always equals *energy out* for the atmosphere. In more technical terms, this means that the heat capacity of the atmosphere is zero. This is not a bad assumption because, while not zero, the heat capacity of the atmosphere is indeed much smaller than the heat capacity of the rest of the climate system.

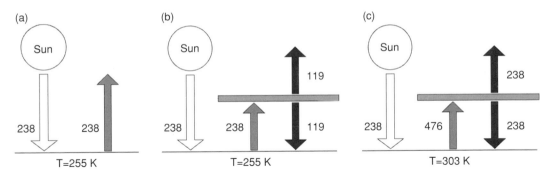

**Figure 6.1** Schematic of energy fluxes on a planet (a) with no atmosphere, (b) the instant after a one-layer atmosphere is added to the planet, and (c) after the climate reaches its new equilibrium.

the energy that is no longer escaping to space, 119 W/m², has been redirected by the atmosphere back toward the surface. As a result, *energy in* for the surface is now $E_{in} = 238 + 119 = 357$ W/m², which exceeds *energy out* of $E_{out} = 238$ W/m², so the surface is also warming. As the surface and rest of the planet warms, *energy out* increases. Eventually, the surface and atmosphere warm enough that $E_{in} = E_{out}$ and the planet is again in energy balance; this is shown in Figure 6.1c.

The scenario just described is basically what occurs when greenhouse gases are added to our atmosphere. The greenhouse gases intercept some of the energy escaping to space and redirect it back toward the surface. In doing so, greenhouse gases both reduce *energy out* for the planet and increase *energy in* for the surface, thereby knocking the system out of equilibrium and forcing the climate system to warm. The planet warms until *energy out* again balances with *energy in* for the planet as a whole and for each individual component of the climate system.

How long does it take for the planet to reach a new equilibrium temperature after the addition of greenhouse gases to the atmosphere? Most of the thermal inertia of the climate system on Earth comes from the ocean: it covers 70 percent of the Earth to an average depth of 4,000 m (2.5 miles). However, you shouldn't think of the ocean as a single entity. Just like in Chapter 5, it is more accurate to consider the ocean split into two parts: a mixed layer, a few hundred meters thick, that communicates rapidly with the atmosphere and the deep ocean containing the other roughly 95 percent of the ocean's mass, which communicates slowly with the atmosphere.

The mixed layer is in contact with the atmosphere, so greenhouse warming directly heats it. And because its mass is relatively small (compared to the mass of the entire ocean), it warms rapidly and reaches equilibrium after just a few decades. After that, the rate of warming of the climate system is controlled by the rate of heating of the deep ocean. This deep-ocean warming is so slow that it takes millennia for the entire ocean to reach equilibrium.

Thus, after release of a slug of carbon dioxide, you get rapid warming for a few decades (as the mixed layer warms) followed by slower warming for thousands of years (as the deep ocean warms). Because of this, the carbon dioxide we are emitting today will still be warming the climate in the year 3000 and beyond. A sobering thought, indeed.

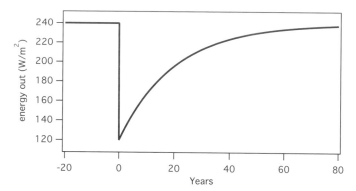

**Figure 6.2**   Plot of *energy out* for the planet shown in Figure 6.1. The atmosphere is added instantaneously in Year 0.

This lag between emission of greenhouse gases and the resultant warming has some important implications. Because we are constantly emitting greenhouse gases, climate change is presently lagging the amount of greenhouse gases in our atmosphere. So if we stopped emitting greenhouse gases today, *the climate would continue to warm for centuries.* Future warming from emissions that have already occurred is likely to be about half a degree Celsius,[2] comparable to the warming of the last century. This is often referred to as *committed warming* or, more informally, as warming "in the pipeline." We have already paid for this warming with emissions over the past few decades, but the warming has not yet arrived. Nevertheless, it is coming – and there is little we can do to avoid it.

The lag between emissions and warming also has important implications for the policy debate over climate change. Because it takes decades to experience the bulk of the warming from today's emissions, the benefits from reductions in emissions today, which may be costly, will also not be fully realized for many decades. This means that reducing emissions requires today's society to pay costs that may primarily benefit future generations. This creates an incentive for policymakers to do nothing about climate change – "kick the can down the road," as they say – and leave the problem for people in the future to solve. When we explore policy options later in the book, we will return to this problem.

## 6.2 Radiative forcing

Figure 6.2 plots *energy out* to space for the planet shown in Figure 6.1. The period before Year 0 corresponds to Figure 6.1a, when the planet had no atmosphere and it was in energy balance. At Year 0, the atmosphere is instantaneously added, and *energy out* drops immediately to 119 W/m². This corresponds to Figure 6.1b. As the surface heats up in response to the warming from the atmosphere, *energy out*

---

[2]   The exact amount of warming depends on what we assume for emissions of non-greenhouse-gas forcers, such as that from aerosols.

increases, eventually reaching 238 W/m$^2$ – and the planet is once again in energy balance, corresponding to Figure 6.1c.

This leads us to one of the most important concepts in climate science: *radiative forcing*. Radiative forcing (often abbreviated RF) is the change in $E_{in} - E_{out}$ for the planet as a result of some change imposed on the planet *before the temperature of the planet has adjusted in response*[3]:

$$\mathrm{RF} = \Delta \left( E_{in} - E_{out} \right) = \Delta E_{in} - \Delta E_{out} \qquad (6.1)$$

In the example just given, $\Delta E_{out}$ is $-120$ W/m$^2$; that is the drop in $E_{out}$ the instant after the atmosphere is added but before the warming temperature has caused $E_{out}$ to increase. Note that $\Delta E_{in}$ is zero because *energy in* does not change when an atmospheric layer is added. Thus, the radiative forcing of adding a one-layer atmosphere is $0 - (-120) = +120$ W/m$^2$. The sign convention is that positive radiative forcings correspond to changes that warm the climate, whereas negative ones correspond to changes that cool the climate.

---

**An example: What is the radiative forcing of a 5 percent increase in solar constant for the Earth that occurs over 100 years?**

Let us begin by calculating $\Delta E_{in}$, the change in *energy in*. From Chapter 4, we know that $E_{in} = S(1 - \alpha)/4$, which for the Earth is 238 W/m$^2$. If the solar constant $S$ increased by 5 percent, then $S$ would increase to $1,360(1.05) = 1428$ W/m$^2$. For this new value of the solar constant, $E_{in} = 250$ W/m$^2$. Thus, $E_{in}$ has increased from 238 W/m$^2$ to 250 W/m$^2$, so $\Delta E_{in} = +12$ W/m$^2$.

What is $\Delta E_{out}$ for this solar constant change? $E_{out}$ is determined entirely by atmospheric composition (i.e., number of layers) and temperature. Atmospheric composition is not changing in this example, and radiative forcing is defined as the response to an instantaneous change, *before the temperature of the planet has adjusted to the change*. Thus, $E_{out}$ does not change and $\Delta E_{out} = 0$. Putting it together using Equation 6.2, we see that the radiative forcing for this change in the solar constant is $+12$ W/m$^2$.

The fact that the change occurred over the course of 100 years does not enter into the calculation. Radiative forcing calculations are done under the assumption that the climate is not allowed to respond to the change, so the length of time the change is imposed over is irrelevant.

---

The imposition of a radiative forcing on a planet, such as a change in solar constant, will take the planet out of energy balance – so that $E_{in}$ and $E_{out}$ are no longer equal to each other. In response, the temperature of the planet will adjust so that $E_{in}$ once again equals $E_{out}$. In the case of an increasing solar constant, the planet will warm, increasing $E_{out}$, until $E_{in}$ and $E_{out}$ are once again in balance.

Thus, radiative forcing is a quantitative measure of how much some climate perturbation (e.g., an increase in solar constant, increase in greenhouse gases) will change the climate. The advantage of using radiative forcing is that it allows us to express

---

[3] For somewhat technical reasons, radiative forcing is usually defined as the change in energy balance at the tropopause. We ignore this technical point here.

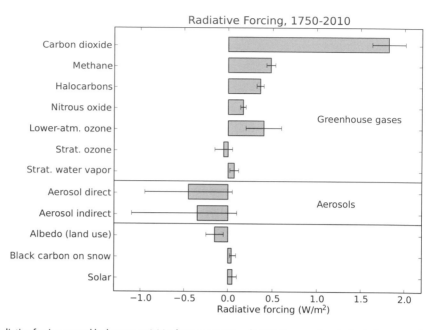

**Figure 6.3** Radiative forcing caused by human activities between 1750 and 2010. The error bars indicate the uncertainty of the estimate (based on values from Ciais et al., 2013).

diverse changes to the climate system by using a common metric. For example, it allows us to compare the climate-changing effect of a 100-ppm increase in carbon dioxide to a 1 percent increase in the solar constant. By comparing the radiative forcing of these two changes, we could determine which one would warm the planet more. Similarly, radiative forcing of $+1$ W/m$^2$ will produce similar warming of the climate, regardless of whether that change was caused by a brightening of the Sun, an increase in carbon dioxide, an increase in methane, or some other change.

Figure 6.3 shows the radiative forcing of the various factors that have influenced our climate over the past few centuries. In much the same way that temperature anomalies are the change in temperature from a reference period, radiative forcings are generally calculated as a change from a reference climate. In Figure 6.3, the values plotted are radiative forcing caused by changes since 1750, which is considered the preindustrial value. In the rest of this section, I describe each one of these factors.

## 6.2.1 Greenhouse gases

The atmospheric abundance of carbon dioxide increased from 280 to 391 ppm between 1750 and 2010; this change reduces *energy out* by 1.82 W/m$^2$ (*energy in* does not significantly change). Thus, the radiative forcing of carbon dioxide is $+1.82$ W/m$^2$. Increases in methane, nitrous oxide, and the halocarbons between 1750 and 2010 produced radiative forcings of $+0.48$, $+0.16$, and $+0.36$ W/m$^2$, respectively, for a total of $+1.00$ W/m$^2$.

Ozone in the lower atmosphere is both a greenhouse gas and one of the primary components of photochemical smog. As the world has become more industrialized, lower atmospheric ozone has increased along with the other components of air pollution. This increase contributes a positive radiative forcing of $+0.4$ W/m$^2$. Ozone in the stratosphere, in contrast, has been declining as a result of ozone depletion from halocarbons and other manmade chemicals. This contributes a negative radiative forcing of $-0.05$ W/m$^2$.

Next is stratospheric water vapor. An important source of stratospheric water vapor is the transport of methane into the stratosphere followed by oxidation, which has this net reaction:

$$CH_4 + 2O_2 \rightarrow 2H_2O + CO_2 \qquad (6.2)$$

The increase in methane over the past two centuries has therefore increased stratospheric water, which has led to a positive radiative forcing of $+0.07$ W/m$^2$. We will consider lower-atmospheric water vapor in the section on feedbacks.

Thus, although carbon dioxide is the single most important greenhouse gas emitted by human activities, it is not the only important one. In fact, the combined radiative forcing from the other greenhouse gases ($+1.42$ W/m$^2$) is nearly as large as the radiative forcing from carbon dioxide alone ($+1.82$ W/m$^2$). This has important implications for policies to address climate change, as we will discuss later.

## 6.2.2 Aerosols

Aerosols are particles so small that they do not fall under the force of gravity but remain suspended in the atmosphere for days or weeks. Aerosols can interact both with sunlight that is falling on the planet and with infrared radiation that is being emitted by the surface and atmosphere – thereby altering the climate. There are several types of aerosols, and their composition determines how they interact with sunlight and infrared radiation.

When fossil fuels containing sulfur impurities are burned, the sulfur is released to the atmosphere with the other products of combustion. Sulfur is also released into the atmosphere during biomass burning and from natural processes in the ocean. Once in the atmosphere, the sulfur gases react with other atmospheric constituents to form small liquid droplets, known as sulfate aerosols.

Sulfate aerosols are highly reflective and reflect incoming solar radiation back to space, so their net effect is to cool the climate. As a result of increases in fossil fuel use over the past two centuries, the abundance of sulfate aerosols has steadily increased with time, providing a negative radiative forcing of $-0.4$ W/m$^2$.

Such sulfate aerosols occur in the lower atmosphere, so their lifetime is short – it takes just a few weeks before the aerosols are either washed out of the atmosphere by rain or fall to the ground. This means that the radiative forcing the Earth is experiencing at any given time from these aerosols is due entirely to emissions of sulfur that occurred in the past month or two.

Another important – but episodic – source of sulfur gases for the atmosphere is volcanic eruptions. Volcanoes emit enormous amounts of sulfur gas, and energetic eruptions can inject it directly into the stratosphere. Aerosols in the stratosphere can remain there for several years – much longer than an aerosol resides in the lower atmosphere. This long lifetime, combined with the massive amounts of

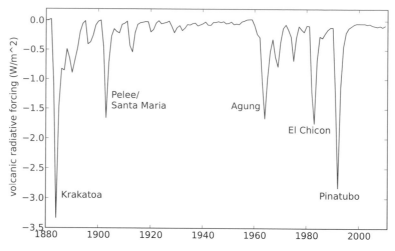

**Figure 6.4**  Radiative forcing from volcanoes (obtained from NASA GISS; see data.giss.nasa.gov/modelforce/Fe.1880–2011.txt).

sulfur released, means that a single volcano can produce a negative radiative forcing of several watts per square meter that lasts for several years after the eruption (Figure 6.4).

This negative radiative forcing can lead to a noticeable cooling of the climate following an eruption. In 1816, for example, after three major eruptions in three years, the United States and Europe experienced the "year without a summer," in which snow fell in Vermont in June and heavy summer frosts caused crop failures and widespread food shortages. When that summer was followed by a winter so cold that the mercury in thermometers froze (this happens at $-40\,^{\circ}C$), many residents fled the Northeast United States and moved south.

A few years after a volcanic eruption, volcanic aerosols fall out of the stratosphere and the climate warms back up. Combined with the fact that such massive volcanic eruptions occur infrequently (as we can see in Figure 6.4), the long-term impact of volcanoes on the climate has been relatively small over the past few centuries.

Black carbon aerosols, such as soot, are another important aerosol type. This type of aerosol is produced by incomplete combustion, such as a smoldering fire or by two-stroke gasoline engines, so they are generally of human origin. Because they are dark, they absorb solar radiation and decrease the planet's albedo, thereby warming the planet. Over the past few centuries, black carbon aerosol abundance has increased as more people burn more stuff, leading to a positive radiative forcing of $+0.4\,W/m^2$. Much like sulfate aerosols, these black carbon aerosols have atmospheric lifetimes of a few weeks.

Another type of aerosol is mineral dust. Most of this dust comes from natural processes, such as dust picked up off the world's deserts by strong winds (Figure 6.5). But approximately 20 percent of mineral dust comes from anthropogenic sources – mainly agricultural practices (e.g., harvesting, plowing, overgrazing), changes in surface water features (e.g., drying out of lakes such as the Aral Sea and Owens Lake) and industrial practices (e.g., cement production, transport). The net effect of dust is

**Figure 6.5**    Image of a strong temperate cyclone over China, pushing a wall of dust as it moved. The image was captured in early April of 2001 by the Moderate Resolution Imaging Spectroradiometer on NASA's Terra satellite (image obtained from the Earth Observatory; see earthobservatory.nasa.gov/IOTD/view.php?id=8341). (See Color Plate 6.5.)

to cool the planet, and human activities have contributed a negative radiative forcing of $-0.1$ W/m$^2$. Like other types of aerosols, these dust aerosols have atmospheric lifetimes of a few weeks.

Combining all types of aerosols (those discussed earlier and several not discussed), the direct radiative effect of aerosols is to cool the climate, with an estimated negative radiative forcing of $-0.35$ W/m$^2$. However, aerosols also have an indirect effect on the climate, whereby aerosols influence the climate by altering clouds. There are several ways that this can occur. One of the clearest mechanisms is by altering the number of particles in a cloud. Cloud particles generally form when water condenses onto *cloud-condensation nuclei* or CCN, which are small solid or liquid aerosols that are hydrophilic, meaning that they attract water. It is the number of CCN in an air mass that determines how many cloud particles are found in a cloud.

If you add aerosols to a cloud, then you will increase the number of CCN – and therefore the number of droplets making up the cloud. But the total liquid water contained in the cloud is fixed. Thus, the increase in the number of droplets means that each droplet has less water and is therefore smaller. This is akin to cutting a pie

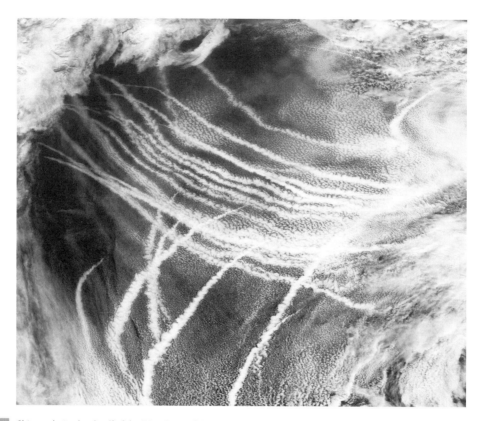

**Figure 6.6**    Ship tracks in clouds off of the West Coast of the United States (image obtained from the Earth Observatory, see earthobservatory.nasa.gov/IOTD/view.php?id=37455).

into more slices: The total amount of pie is fixed, so more slices means that each slice must be smaller.

It turns out that a cloud containing smaller droplets is more reflective than one containing large droplets. A familiar example of this can be seen in your kitchen in the difference between regular table sugar and powdered sugar. Chemically, the two substances are identical, but powdered sugar is made up of much smaller particles. Because smaller particles tend to be more reflective, the pile of powdered sugar is a brighter white than a pile of table sugar.

Thus, if one adds aerosols to a cloud, the cloud gets brighter and more reflective. This can be seen in what are called *ship tracks*. The exhaust from diesel engines contains fine particulates that can serve as CCN. As ships steam across the ocean, these fine aerosol particles from their engines are transported by the winds into low-level clouds, leading to increases in numbers of droplets and brighter clouds. From a satellite (Figure 6.6), lines of bright clouds trace out the paths of these ships.

This effect on cloud reflectivity is just one effect aerosols have on clouds. By making the cloud particles smaller, aerosols also slow down the coagulation process whereby cloud droplets combine to form raindrops. This reduces rainfall from a cloud, so the clouds last longer. Aerosols can also change the height of the cloud, as

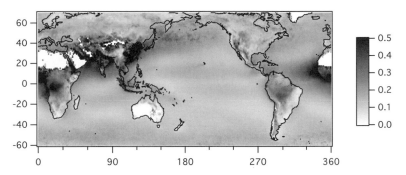

**Figure 6.7** Annual average aerosol optical depth (a measure of the abundance of aerosols) as a function of latitude and longitude, for the years 2004–2008 (measurements were made by the Moderate Resolution Imaging Spectroradiometer onboard NASA's Aqua satellite and were obtained from the NASA Goddard Earth Sciences Data and Information Services Center). White areas are regions where no data were obtained. (See Color Plate 6.7.)

well as the phase (ice vs. liquid). The addition of black carbon to clouds can lead to local warming that can cause clouds to evaporate.

Considering all the effects of aerosols on clouds, scientists estimate that the indirect aerosol effect produces a negative radiative forcing of $-0.45$ W/m$^2$. However, as the error bars in Figure 6.3 show, the indirect effect of aerosols is highly uncertain. Combining the direct and indirect effects, aerosols produce a negative radiative forcing of roughly $-0.9$ W/m$^2$.

Because aerosols last only a few weeks in the atmosphere before they are removed, aerosols do not have time to become well mixed throughout the atmosphere (which takes a year or so). As a result, the distribution of aerosols in our atmosphere is highly variable, with most aerosols found near their sources. Figure 6.7 shows their distribution, and from this we can infer the major sources of aerosols across the globe: Saharan dust that is blown westward from North Africa into the Atlantic, smoke from biomass burning over central Africa, and an aerosol soup that originates over Asia and is blown eastward over the Pacific toward the United States.

It is also apparent that aggressive air-pollution control efforts in the United States and Western Europe have worked – these regions have low levels of aerosol abundance despite large economies. It is the poorer countries with lax environmental regulations, such as China and India, that are responsible for much of the anthropogenic aerosols in the atmosphere.

So although the global average radiative forcing from aerosols is $-0.9$ W/m$^2$, this is not evenly distributed over the globe. In regions where aerosol abundance is high, aerosols can have a local radiative forcing of many times this value, whereas in regions that have no aerosols, the local radiative forcing can be zero. This can be contrasted to greenhouse gases such as carbon dioxide or methane, which are well mixed in the atmosphere because of their long atmospheric lifetimes (many years), resulting in a radiative forcing evenly distributed across the globe.

From a climate perspective, the negative radiative forcing from aerosols offsets 25 percent of the positive radiative forcing from anthropogenic greenhouse gases. In this way, aerosols benefit us, because without them the net radiative forcing would be higher – and global warming would be worse. But aerosols are not all good – they are also one of the main components of air pollution around the world, which kills millions of people every year.

As these poorer countries begin to clean up their atmosphere, we expect to see the amount of aerosols in the atmosphere diminish. Although such reductions improve public health, they also make global warming worse by reducing the cooling that aerosols provide. This is another factor that must be considered in climate change policy.

## 6.2.3 Land-use changes

Humans have been remaking the face of the planet for thousands of years, and over the past few centuries, humans have altered vast swaths of the surface. For example, in 1750, approximately 7 percent of the global land area was under cultivation or pasture; in 1990, that number was slightly more than a third. Such alterations of the surface can modify the Earth's climate. Agricultural land typically has a higher albedo than does the natural landscape, especially if the latter is forest. Thus, cutting down a forest and replacing it with grassland for grazing cattle will increase the surface's albedo. Over all, human *land-use changes* have tended to cool the planet, with a radiative forcing of $-0.15$ W/m$^2$. This albedo effect is in addition to any radiative forcing from carbon emissions from land-use changes, which were discussed in Section 5.4.

Another way human activities can alter the albedo of the surface is through the release of black carbon – basically soot. In the previous section we explored how these dark particles have a warming effect when suspended in the atmosphere because they absorb sunlight that falls on them. However, their climate impact does not end there. After a month or two in the atmosphere, these particles are removed from the atmosphere and deposited on the surface. If the surface is bright, like snow, then this deposition will reduce the albedo of the surface, thereby increasing the absorption of solar energy. With the increase in industrial activities over the past two centuries, the amount of black carbon deposited on snow has led to a positive radiative forcing with an estimated magnitude of $+0.04$ W/m$^2$.

## 6.2.4 Changes in the Sun

A final radiative forcing – and one that is entirely unrelated to humans – comes from changes in the amount of solar energy reaching the Earth. As we will explore in more detail in Chapter 7, rather than being a constant source of energy, the Sun has well-documented variations. The best known is the eleven-year sunspot cycle, over which the solar constant varies by roughly 1 W/m$^2$ (out of 1,360 W/m$^2$, so it is not a terribly big variation). The Sun's output also varies slightly with a period of twenty-seven days, which is the time it takes the Sun to rotate once on its axis. Because of the large heat capacity of the climate system, neither of these variations in the Sun have much effect on the climate.

However, it is possible that, over longer periods, the Sun's output can vary – and this might have an important influence on climate. Unfortunately, our knowledge of these longer-term variations is poor. Accurately measuring the output of the Sun must be done from orbit, and that means that we do not have reliable measurements of the solar constant before the late 1970s, when the first satellites designed to measure the Sun's output were launched. Scientists have, however, attempted to reconstruct the long-term record of the solar constant by using proxy data, such as sunspot number. This work has suggested that the Sun may have gotten slightly brighter over the past 250 years, particularly early in the twentieth century, and provided a positive radiative forcing estimated to be +0.05 W/m$^2$.

## 6.2.5 Total net forcing

Summing all of the radiative forcings in the period between 1750 and 2005, we get a net radiative forcing of +2.3 W/m$^2$. Considering uncertainties (expressed as error bars in Figure 6.3), the total radiative forcing could actually be anywhere between +1.1 and +3.3 W/m$^2$. This total comes from positive forcings, primarily the greenhouse gases, and negative forcings from aerosols and land-use changes.

Let me emphasize what this value means: If we made all of the changes in greenhouse gases, aerosols, land use, and so on instantaneously and without letting the atmosphere warm, $E_{in} - E_{out}$ would have increased 2.3 W/m$^2$. In order for the Earth to once again be in energy balance, the planet would have to warm, which would increase $E_{out}$. Earlier in the chapter, I said that, for the present-day Earth, $E_{in}$ exceeds $E_{out}$ by approximately 0.5 W/m$^2$. This means that the planet has warmed up enough in the past 250 years to erase +1.8 W/m$^2$ of the radiative forcing. The planet needs to continue warming in order to erase the remaining radiative forcing – this is warming that we are already committed to and can do little to stop.

Another important take-away message from this section is that climate change is about much more than carbon dioxide. Carbon dioxide is indeed the most important greenhouse gas that humans are adding to the atmosphere and the most important climate forcing of the past few centuries. But other greenhouse gases that humans are adding are also making important contributions to the radiative forcing. And non-gas constituents, such as aerosols, also play an important role.

You should now be able to see the foundation of the policies we might undertake to stabilize our climate. The temperature of the climate is on an upward trajectory because we are increasing the net radiative forcing through emissions of greenhouse gases. If we stabilize net radiative forcing, then the climate will stabilize a few decades later. There are two obvious ways to do this. First, we can stop activities that produce positive radiative forcing – this basically means ceasing emissions of greenhouse gases to the atmosphere. In the policy world, this is known as mitigation. Alternatively, we could intentionally engage in activities that produce negative radiative forcing – for example, we could add sulfate aerosols to the atmosphere, thereby canceling out the positive radiative forcing. This latter approach is what is commonly referred to as geoengineering. I will have more to say about both approaches in Chapters 11 and 12.

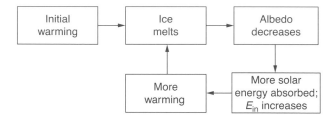

**Figure 6.8**  The ice-albedo feedback loop.

# 6.3 Feedbacks

One of the ultimate goals of climate science is to make quantitative predictions of the future climate. As discussed in Chapter 4, given the solar constant, albedo, and atmospheric composition, it is conceptually easy to calculate the temperature of a planet. You may expect, therefore, that it would be pretty easy to calculate how much warming we should expect if we applied some specified radiative forcing to the planet (e.g., carbon dioxide is doubled, the Sun gets 1 percent brighter).

For example, consider a hypothetical planet with no atmosphere ($n = 0$) and an Earth-like solar constant and albedo ($S = 1,360$ W/m$^2$, $\alpha = 0.3$). The surface temperature of this planet is, according to Equation 4.5, $T = 254.5$ K. Now imagine that greenhouse gases are added to the atmosphere, so that n increases to 0.01. According to Equation 4.5, the new temperature would be 255.2 K, which is a warming of 0.6 K.

That answer would be correct if nothing else in the climate system changed. But as the planet warms, other things do change. For example, because ice melts reliably at $0°C$, global warming should reduce the amount of ice on the Earth's surface – and, as discussed in Section 2.1.3, that is indeed happening. If the melting ice uncovers a dark surface, such as ocean, then this decreases the average planetary albedo (i.e., makes the planet less reflective), which leads to more absorption of solar radiation and additional warming. This additional warming leads to even more melting, which leads to further decreases in albedo and further warming, and so on. This is known as a feedback loop, and it is shown schematically in Figure 6.8.

The net effect of the feedback loop shown in Figure 6.8 is to amplify the initial warming from the addition of greenhouse gas. To provide a quantitative demonstration of this, let us make the reasonable assumption that the albedo is proportional to the planet's surface temperature and that the relation between these two variables is: $\alpha(T) = 0.3 - (T - 254.54)/150$. This relation quantifies the change in albedo as the planet's temperature changes.

The equation for surface temperature now includes an albedo that is a function of $T$:

$$T = \sqrt[4]{\frac{(n+1)\,S\,[1 - \alpha\,(T)]}{4\sigma}} = \sqrt[4]{\frac{(n+1)\,S\,[1 - (0.3 - (T - 254.54)\,/150)]}{4\sigma}} \quad (6.3)$$

This is no longer a trivial equation to solve because $T$ appears on both sides of the equation. Rather, this must be solved by factoring a fourth-order polynomial or by solving the equation numerically.

Using one of those techniques, we can calculate the temperature of a planet with $n = 0$ and $S = 1,360$ W/m$^2$ (we no longer specify albedo because it is a function of temperature). Solving this equation, we find that the temperature is 254.5 K, the same as when we assumed the albedo $\alpha = 0.3$. This is not an accident because we defined the albedo to have a value of 0.3 at that temperature.

Now let us add the same amount of greenhouse gases to the atmosphere as we did in the no-feedback example, which increases $n$ to 0.01. Now the solution to Equation 6.4 is $T = 256.2$ K, which yields a warming of 1.7 K. In other words, the inclusion of melting ice raises the warming from 0.6 K to 1.6 K. This amplification of the warming is referred to as a *positive feedback.*

You can also have a *negative feedback,* which acts to reduce the initial warming. Imagine a planet that is covered with flowers of two colors: white and black. As the temperature of the planet goes up, the white flowers prosper while black flowers die. This means that, as a planet warms, the planet also becomes whiter – i.e., the albedo goes up.

To quantitatively explore the effects of this, let us assume that the relation between albedo and temperature on this flower-covered planet is $\alpha(T) = 0.3 + (T - 254.5)/150$, so that an increase in $T$ leads to an increase in $\alpha$. In this case, an increase in greenhouse gases from $n = 0.0$ to $n = 0.01$ leads to a new surface temperature of 254.9 K, which is a warming of only 0.4 K. This is less warming than we would get if the albedo were constant. Thus, the flowers have reduced the amount of warming and this flower feedback is negative.

## 6.3.1 Fast feedbacks

There are several important feedbacks in our climate system. The feedback just described, in which warmer temperatures melt ice, leading to reduced albedo and additional warming, is known as the ice-albedo feedback. Another important feedback is the water-vapor feedback, which arises because a warmer atmosphere can hold more water vapor. Thus, global warming leads to increased atmospheric humidity, and because water vapor is itself a greenhouse gas, this leads to additional warming. Both the ice-albedo and water-vapor feedbacks are positive, meaning that they amplify an initial warming.

There are also negative feedbacks in the climate system. The biggest negative feedback is known as the lapse-rate feedback. Because power radiated by a blackbody is equal to $\sigma T^4$, a warmer atmosphere radiates more power to space. Therefore, as the upper atmosphere warms, the enhanced radiation offsets some of the initial warming.

The biggest debate among scientists today is about cloud feedback. Clouds affect the climate in two opposite ways. First, they reflect sunlight back to space, reducing *energy in*, which tends to cool the climate. Second, they absorb infrared radiation emitted by the surface, decreasing *energy out* just like a greenhouse gas, and this tends to warm the climate. The net effect of clouds on the climate is therefore the difference of these two opposing effects. In our present climate, the reflection of solar radiation is slightly larger than the heat trapping effects, so clouds reduce $E_{in} - E_{out}$ for the Earth by roughly 25 W/m$^2$.

This could, however, change in the future. If, in response to an initial warming, the cooling effect of clouds is enhanced, then the effect of clouds on the planet's energy budget will be more negative and clouds will act to reduce the initial warming and therefore be a negative feedback. In contrast, if the heat trapping is enhanced, then the cloud's radiative effect will become less negative and clouds will amplify the initial warming and therefore be a positive feedback. Although the exact answer is uncertain, the best guess of the scientific community is that the climate feedback is positive, meaning that clouds amplify warming.

Feedbacks discussed in this section are known as fast feedbacks because they occur rapidly enough in response to a change in surface temperature that they will play an important role in the evolution of climate change over the coming century. Water vapor, clouds, and the lapse rate all respond within a week or so to changes in surface temperature, and therefore their impact on *energy in* and *energy out* is fast. The response time of ice depends on the type. Some types (e.g., snow) will respond in weeks or months while others (e.g., sea ice) will respond in years, so changes in these types could also be considered a fast feedback.

## 6.3.2 Slow feedbacks

In contrast to fast feedbacks, slow feedbacks include processes that respond slowly to increasing surface temperature, so they require long periods of warmth before they significantly alter *energy in* or *energy out*. For example, the Greenland and Antarctic ice sheets are so big that they will likely take many centuries to significantly respond to a change in temperature. The ice-albedo feedback associated with ice sheets would therefore be categorized as a slow feedback.

Another slow feedback revolves around the fact that there are large amounts of carbon stored in the ground. One of these carbon reservoirs is permafrost, which was discussed in Section 5.2.1. This permafrost contains dead organic plant matter that is kept intact as long as the ground remains frozen. If a warming climate leads to the melting of permafrost, then the organic matter in it thaws out and decays, releasing the carbon back into the atmosphere in the form of either methane or carbon dioxide.

We know that permafrost is indeed melting, which is consistent with the large warming in the Arctic over the past few decades (e.g., see Figure 2.2b). In Alaska, for example, roads and buildings constructed on permafrost under the assumption that the permafrost would never melt are now suffering damage as the permafrost melts and the ground underneath begins shifting. In Siberia, melting permafrost formed during the last ice age is revealing frozen oddities such as intact woolly mammoths that died 20,000 years ago.

Another source of frozen greenhouse gases is what are known as methane clathrates – methane molecules that are embedded in ice. Clathrates can exist on land or under the ocean, and as with permafrost, warming temperatures can melt the ice and release the methane trapped therein. Given that a molecule of methane is twenty times as powerful a greenhouse gas as a molecule of carbon dioxide, the rapid release of significant amounts of methane is a worrying possibility. And there are several other reservoirs of carbon, such as the tropical forests, which may release the carbon to the atmosphere as the climate warms.

This opens the possibility of a carbon-cycle feedback, in which an initial warming leads to the release of large amounts of carbon dioxide and methane that are currently frozen in the ground or otherwise sequestered. The release of these greenhouse gases leads to more warming, and the further release of greenhouse gases. The occurrence of such a feedback in the next few centuries is speculative, but there is reasonably strong evidence that they have occurred in the past, such as during ice-age cycles. I will return to this point when I discuss ice ages in Chapter 7.

Another slow feedback involves vegetation. It has long been known that the distribution of vegetation on the Earth's surface is governed to a large extent by the climate – through the distribution of precipitation, temperature, sunlight, and other such factors. Recently, however, it has been realized that changes in vegetation can also affect the climate. For example, the conversion of a forest to grassland will increase the albedo (because the forest is darker than the grassland), thereby tending to cool the climate. Changes in vegetation can also directly impact exchanges of heat, water, and momentum between the surface and atmosphere, or modify the rate of uptake of carbon dioxide by the vegetation. This introduces the possibility of vegetation feedbacks in which changes in climate lead to changes in vegetation, which in turn lead to additional changes in climate.

Probably the slowest feedback is the weathering thermostat. As the Earth's surface warms, the total amount of rainfall will also increase. The increase in rainfall in turn increases the rate of chemical weathering, which removes carbon dioxide from the atmosphere (we explored this in Section 5.3). The reduction in atmospheric carbon dioxide acts to offset some of the initial warming, so this is a negative feedback that tends to stabilize the Earth's temperature. That is the good news. The bad news is that the weathering thermostat operates on geologic time scales, so it only has an impact on the climate over millions of years. We should not expect the weathering thermostat to ameliorate warming over the next century – or over any time period that human society cares about.

In general, slow feedbacks are much more uncertain than fast feedbacks because they are so slow that modern Earth science, which is really only a few decades old, simply does not have data extending over periods long enough to observe, understand, and quantify them. Thus, the net effect of slow feedbacks on the climate is uncertain. Nevertheless, they continue to compel our attention because many of the worst-case climate scenarios involve slow feedbacks causing extremely large warming in the next few centuries.

## An aside: Feedback vs. radiative forcing

It is worth explicitly discussing the differences between climate feedbacks and radiative forcings. Feedbacks are processes that respond to changes in the Earth's surface temperature, so feedbacks do not initiate climate change. Rather, positive feedbacks amplify and negative feedbacks ameliorate an initial warming. Water vapor, for example, is considered a feedback because the amount of water vapor in the atmosphere is set by the surface temperature of the Earth. If the surface temperature increases, then the amount of water in the atmosphere will also increase, leading to additional warming.

Radiative forcings, in contrast, affect the climate but are themselves unaffected by the climate. The changes in carbon dioxide, methane, and the like between 1750 and 2005 are fundamentally unrelated to the Earth's temperature; instead they are driven by economic activities of human society. Ditto for aerosols.

A confusion arises because some things can be both feedbacks and forcings. For example, although carbon dioxide has been a forcing over the past two centuries, it can also be a feedback if warming temperatures lead to the release of carbon dioxide. Changes in vegetation are a forcing when humans are modifying the vegetation, but they are a feedback when it is the climate that causes the modification.

In most cases, it is clear whether a process is a forcing or feedback. The increase in carbon dioxide over the last two centuries is clearly driven by human activities, not surface temperature, so it is certainly a forcing. Other processes are more ambiguous. For example, the processes that regulate stratospheric water vapor are not well understood. As a result, we do not know if changes in stratospheric water vapor that are now occurring are due directly to human activities, so they would be a forcing, or indirectly, through changing surface temperature, in which case they would be a feedback. In this chapter, we put it in the forcing category, but more subsequent scientific research may reveal that it belongs in the feedback category. Or perhaps it belongs in both categories.

# 6.4 Climate sensitivity

## 6.4.1 Feedback math

To get more quantitative about feedbacks, let us first go over some basic feedback math. Consider a feedback loop, such as the ice-albedo feedback (Figure 6.8). We will express the strength of this feedback as $g$, which is the additional fractional warming produced by one trip through the feedback loop per degree of initial warming. Thus, in response to an initial warming $\Delta T_i$, the first trip through the feedback loop produces additional warming of $g\Delta T_i$. But the feedback also operates on this additional warming $g\Delta T_i$, and this produces an additional warming of $g(g\Delta T_i) = g^2 \Delta T_i$. And feedbacks operate on this additional warming too, leading to an additional warming of $g^3 \Delta T_i$, etc. This goes on forever, so the final warming $\Delta T_f$ is:

$$\Delta T_f = \Delta T_i + g\Delta T_i + g^2 \Delta T_i + g^3 \Delta T_i + g^4 \Delta T_i + \cdots \qquad (6.4)$$

We can write this more compactly as

$$\Delta T_f = \sum_{k=0}^{\infty} g^k \Delta T_i \qquad (6.5)$$

The math ninjas among you will recognize that this infinite series can be rewritten more simply as

$$\Delta T_f = \frac{\Delta T_i}{(1-g)} \tag{6.6}$$

If $g = 0$, then there is no feedback and the final temperature change is equal to the initial temperature change. If $g$ is between 0 and 1, then $\Delta T_f$ is larger than $\Delta T_i$, meaning the net sum of feedbacks is positive. If $g$ is less than 0, then $\Delta T_f$ is less than $\Delta T_i$, meaning the net sum of feedbacks is negative.

As positive feedbacks get stronger and $g$ approaches 1, the denominator in Equation 6.7 approaches 0 and $\Delta T_f$ approaches infinity. This should make sense from visual inspection of Equation 6.5: If $g = 1$, then each subsequent term is as big as the previous one and, because the series is infinite, the sum must also be infinite. This situation is sometimes referred to as a "runaway greenhouse effect." In that case, an initial temperature perturbation leads to a very, very large temperature rise.

In Equations 6.5–6.7, $g$ is the sum of the feedback parameters from the individual feedbacks:

$$g = g_{ia} + g_{wv} + g_{cloud} + g_{lr} \tag{6.7}$$

where $g_{ia}$ is the ice-albedo feedback, $g_{wv}$ is the water-vapor feedback, $g_{cloud}$ is the cloud feedback, and $g_{lr}$ is the lapse-rate feedback (we consider here only the fast feedbacks).

The strongest feedback is the water-vapor feedback, with a magnitude $g_{wv} = 0.6$. This feedback is big enough that, by itself, it would more than double the initial warming $\Delta T_i$. The ice-albedo feedback is substantially weaker, with a magnitude $g_{ia} = 0.1$. Because it is a negative feedback, the lapse-rate feedback has a negative magnitude $g_{lr} = -0.3$. Finally, the cloud feedback is quite uncertain, but most scientists would put its magnitude $g_{cloud} = 0.0$–0.3.

Summing these individual feedbacks, we get a total feedback parameter for our climate of $g = 0.4$–0.7. This means that, for our climate, $\Delta T_f = 1.6$–3.3 $\Delta T_i$. Thus, as much as two-thirds of the warming we experience comes from feedbacks rather than the direct heating from greenhouse gases. This is why feedbacks occupy much of the scientific debate over climate change.

## 6.4.2 Sensitivity

For historical reasons, climate sensitivity is most frequently expressed as the warming that occurs if the carbon dioxide is instantaneously increased from 280 ppm, the preindustrial value, to 560 ppm, twice the preindustrial value, and then one lets the climate reach a new equilibrium.

For this doubling of carbon dioxide, the initial warming $\Delta T_i$ is 1.2°C. Using the feedback strengths from the previous section implies a range of final temperature $\Delta T_f = 2$–4°C. A more sophisticated analysis by the IPCC concludes that the climate sensitivity is likely in the range 1.5°C to 4.5°C.

Doubled carbon dioxide corresponds to a radiative forcing of roughly $+4$ W/m$^2$. Given this, we can also express the sensitivity as the warming per unit of radiative forcing. The climate sensitivity in these units is $0.38$–$1.1\,^\circ$C/(W/m$^2$). If you need a single value for the climate sensitivity, I use $0.75\,^\circ$C/(W/m$^2$), which is a convenient value that falls near the middle of the range.

As an example, in Section 6.2 we calculated that the radiative forcing for a 5 percent increase in solar constant is $+12$ W/m$^2$. Given that, we can calculate how much warming we get by multiplying this radiative forcing by the climate sensitivity of $0.75\,^\circ$C/(W/m$^2$), yielding a warming of $9\,^\circ$C.

These values of the climate sensitivity only include the fast feedbacks. This is probably appropriate for climate change over the next century. Over several centuries or longer, though, the contribution of slow feedbacks can become important. These feedbacks are thought to be mainly positive, so the climate sensitivity may be significantly higher when we consider such longer periods. Exactly how much higher is unknown, but looking at previous long-term warming events in response to greenhouse gas emissions (such as the Paleocene-Eocene Thermal Maximum, or PETM, mentioned in Section 2.2.2) provides some evidence that slow feedbacks may increase the climate sensitivity. That would be very bad news for our descendants living several hundred years from now.

## 6.5 Chapter summary

- A radiative forcing is an imposed change on the energy balance of the Earth. It is calculated as the change in energy balance for the planet (*energy in* minus *energy out*) after the imposition of the specific change in the climate but before the climate has changed in response.
- In response to a radiative forcing, the Earth's temperature adjusts so that energy balance is reestablished.
- Because of the Earth's thermal mass, this adjustment happens over time. There is a relatively rapid warming during the first few decades as the ocean's mixed layer warms. After that, it is the warming of the deep ocean that sets the pace of warming. Given the enormous heat capacity of the deep ocean, this warming takes millennia. Thus, the climate will still be warming in 1,000 years as a result of greenhouse gases we emit today.
- The increase in greenhouse gases since 1750 has imposed a radiative forcing of $+3.2$ W/m$^2$. The increase in carbon dioxide is responsible for $+1.8$ W/m$^2$, or more than half of the total forcing. The change in aerosols since 1750 has imposed a net radiative forcing of $-0.9$ W/m$^2$. This means that aerosols offset approximately 25 percent of the radiative forcing from increasing greenhouse gases. Summing all changes, we get a net radiative forcing over this time period of $+2.3$ W/m$^2$.
- Positive feedbacks amplify and negative feedbacks ameliorate an initial warming. For the problem of modern climate change, we are mainly concerned with the following fast feedbacks: water vapor, ice-albedo, lapse rate, and clouds. Together, they double to triple an initial warming.

- Feedbacks are processes that respond to changes in the surface temperature, whereas forcings are unrelated to the surface temperature. Thus, feedbacks do not initiate climate change, but forcings do.
- The Earth's climate sensitivity, which is conventionally defined as the equilibrium temperature increase caused by a doubling of carbon dioxide from 280 ppm to 560 ppm, is 1.5–4.5 °C. In terms of radiative forcing, the climate sensitivity is $0.38–1.1 °C/(W/m^2)$, a range centered on $0.75 °C/(W/m^2)$.

## Additional reading

Myhre, G., D. Shindell, F.-M. Bréon, W. Collins, J. Fuglestvedt, J. Huang, D. Koch, J.-F. Lamarque, D. Lee, B. Mendoza, T. Nakajima, A. Robock, G. Stephens, T. Takemura, and H. Zhang, "Anthropogenic and Natural Radiative Forcing," in T. F. Stocker, D. Qin, G.-K. Plattner, M. Tignor, S. K. Allen, J. Boschung, A. Nauels, Y. Xia, V. Bex, and P. M. Midgley (eds.), *Climate Change 2013: The Physical Science Basis*. Contribution of Working Group I to the Fifth Assessment Report of the Intergovernmental Panel on Climate Change (Cambridge and New York: Cambridge University Press, 2013). This is the IPCC's latest summary of what we know about radiative forcing (download at www.ipcc.ch). It is long, and not always easy to read, but it *is* comprehensive.

Boucher, O., D. Randall, P. Artaxo, C. Bretherton, G. Feingold, P. Forster, V.-M. Kerminen, Y. Kondo, H. Liao, U. Lohmann, P. Rasch, S.K. Satheesh, S. Sherwood, B. Stevens, and X.Y. Zhang, "Clouds and Aerosols," in T. F. Stocker, D. Qin, G.-K. Plattner, M. Tignor, S. K. Allen, J. Boschung, A. Nauels, Y. Xia, V. Bex, and P. M. Midgley (eds.), *Climate Change 2013: The Physical Science Basis*. Contribution of Working Group I to the Fifth Assessment Report of the Intergovernmental Panel on Climate Change (Cambridge and New York: Cambridge University Press, 2013). This is the IPCC's latest summary about feedbacks, clouds, and aerosols (download at www.ipcc.ch). Among other things, it contains a thorough summary of how clouds and aerosols interact.

See www.andrewdessler.com/chapter6 for additional resources for this chapter and www.andrewdessler.com/computer for computer exercises that illustrate some of the important concepts, including the ocean's role in regulating the pace of climate change.

## Terms

Carbon-cycle feedback
Climate sensitivity
Cloud-condensation nuclei
Cloud feedback
Committed warming
Direct radiative effect of aerosols

Fast feedbacks
Feedback
Ice-albedo feedback
Indirect effect of aerosols
Lapse-rate feedback
Radiative forcing
Ship tracks
Slow feedbacks
Thermal inertia
Vegetation feedback
Water-vapor feedback

# Problems

1. When considering how long it takes for radiative forcing to warm up the planet, it is useful to think about the heat capacity of the climate system, which tells us how many Joules are required to raise the temperature by one degree Kelvin.

   The heat capacity of the climate system comes mainly from the ocean, and we can estimate it to be $6 \times 10^{24}$ J/K (this means that, if you add $6 \times 10^{24}$ J to the Earth system, the climate will warm by 1 K).

   a. Imagine you impose a radiative forcing of $+2.3$ W/m$^2$, what rate of warming will result? Express your answer in °C/century.

   b. The actual rate of warming of the planet's surface is predicted to be faster than this over the next century. Why?

2. Imagine a planet where $S = 1360$ W/m$^2$ and $\alpha = 0.3$.

   a. If $n = 0$, what is the temperature of this planet?

   b. If $S$ increases to 1370 W/m$^2$ and $n$ remains equal to 0, what is the new temperature?

   c. Let us include a water vapor feedback: $S$ increases to 1370 W/m$^2$, but the number of layers, $n$, is a function of surface $T$: $n(T) = (T - 254.5)/100$. What is the new surface temperature?

   d. Using the answers to b and c and assuming that no other feedbacks are operating, what is the value of $g$ for this climate?

3. List the important fast feedbacks operating in our climate. Identify whether each is positive or negative.

4. Define climate sensitivity. What is the currently accepted value for our climate?

5. Imagine that we add some carbon dioxide to the atmosphere and the Earth warms by 1 °C. How much warming would there have been if there were no feedbacks?

6. Imagine that our Sun brightens by 1 percent instantaneously.

   a) How long would it take for the Earth to reach its new equilibrium temperature? Is this longer or shorter than the time it would take Mars or Mercury to reach their respective equilibrium temperatures?

   b) What radiative forcing does this change correspond to?

   c) Approximately how much warming would this brightening eventually cause?

d) How would the calculated radiative forcing change if the brightening takes place over 1,000 years instead of instantaneously?

7. Explain why water-vapor changes are considered a feedback and not a forcing.

8. The albedo changes from 0.3 to 0.31 on the Earth. What is the radiative forcing associated with this change?

9. If doubled carbon dioxide has a radiative forcing of 4 $W/m^2$, how much of a change in albedo is required to completely cancel a doubling of carbon dioxide on the Earth (put another way, how much of a change in albedo is required to generate a radiative forcing of $-4W/m^2$)?

10. Imagine that, in addition to the fast feedbacks discussed herein, there was also a fast negative "flower" feedback like that described in this chapter (as the planet warms, white flowers prosper while black ones die out), and that it had a magnitude $g = -0.3$. Estimate the Earth's climate sensitivity.

11. Assume that the Earth has warmed by 5°C since the last ice age, and the change in radiative forcing over that time was $+6.7$ $W/m^2$. On this basis, calculate the climate sensitivity.

    a) Express the climate sensitivity in °C/($W/m^2$).

    b) Express the climate sensitivity in degrees Celsius per doubled $CO_2$.

12. In the northern hemisphere, $E_{in}$ maximizes on June 21, when the sun is most directly overhead. You might therefore expect temperatures to be highest on that day. But for the U.S. Gulf Coast, temperatures do not reach their hottest temperatures until mid-August – several months after the maximum in $E_{in}$. Why?

# Why is the climate changing?

In Chapter 2, we detailed the overwhelming evidence that the Earth's climate is changing – evidence so overwhelming, in fact, that few dispute this anymore. Instead, the most heated argument is over the cause of the warming: Is it caused by human activity, or is it natural? In this chapter, we address this question.

Our strategy here is to examine the mechanisms that have changed climate in the past and test each of them to determine if they could be the cause of the recent warming. You will see that a careful assessment of all of the possible causes yields the conclusion that the most likely explanation for the recent warming is the increase in greenhouse gases in our atmosphere, which we learned in Chapter 5 is due to human activity.

## 7.1 The first suspect: Continental drift

As you probably know, the Earth's continents are moving. Not fast, mind you – they move at about the rate that your fingernails grow – but over tens of millions of years, this *continental drift* can substantially alter the arrangement of the continents across the Earth's surface. Such changes can lead to large changes in the climate through several mechanisms.

For example, the location of continents determines whether ice sheets form. The most important requirement for growth of an ice sheet is summer temperatures cool enough that snow falling during the winter does not melt during the following summer. This is most favorable for land at high latitudes, which get the least sunlight.

Ice sheets matter to the climate because they reflect sunlight, so the formation of an ice sheet increases planetary albedo, thereby increasing the reflection of solar radiation back to space and cooling the planet. And the loss of an ice sheet will warm the climate through the same mechanism. In addition, the location of the continents determines the ocean circulation. The oceans carry huge amounts of heat from the tropics to the polar regions, so changing the ocean circulation can therefore alter the relative temperatures of the tropics and polar regions. A good example of this happened 30 million years ago when the Antarctic Peninsula separated from the southern tip of South America, opening the Drake Passage. This isolated Antarctica and allowed winds and water to flow unhindered around it. This intense circumpolar flow reduced the transport of warm water and air from the tropics to the South Polar region, cooling the Antarctic and helping build the Antarctic ice sheet.

Continental drift can also indirectly affect the climate by regulating atmospheric carbon dioxide. As I discussed in Chapter 5, carbon dioxide is slowly removed from the atmosphere by chemical weathering, which occurs when atmospheric carbon dioxide dissolves in rainwater and then reacts with sedimentary rocks. The movement of the continents can change the pattern of rainfall and expose new rock to the atmosphere, changing the locations and rate of chemical weathering – thereby altering the amount of carbon dioxide in the atmosphere. For example, 40 million years ago the Indian subcontinent collided with Asia, forming the Himalayas and the adjacent Tibetan Plateau. This change in surface topology led to changing wind patterns, bringing heavy rainfall onto a vast expanse of newly exposed rock in these features. The resultant chemical weathering drew down atmospheric carbon dioxide over a period of tens of millions of years. This is one of the reasons the planet has been generally cooling over the past 40 million years (as shown in Figure 2.11)

Thus, continental drift can indeed affect the climate. But could it be responsible for the rapid warming of the past few decades? The answer is no. Because the movement is so slow, it takes millions of years for continental movement to cause significant climate change. Continental movements simply cannot significantly modify the climate over decades or centuries.

# 7.2 The Sun

As we explored in Chapter 4, one of the factors that controls our climate is the solar constant $S$. So if the Sun brightens or dims, we expect the climate to respond by warming or cooling. It is therefore reasonable to ask if the recent warming of the climate can be explained by an increase in the brightness of the Sun.

It is well known that the Sun's output varies on many time scales. For example, solar physicists believe that, over the Sun's 5-billion-year life, as the Sun burned hydrogen and produced helium, the rate of fusion in the Sun has increased as the buildup of helium increased in the density of the Sun's core. This has caused the Sun to become about 30 percent brighter over this time.

Since the late 1970s, instruments on satellites have been accurately measuring the solar constant; the measurements are plotted in Figure 7.1. Over this period, there is a clear eleven-year cycle, during which the solar constant varies by about 0.1 percent (about 1.3 $W/m^2$). This cycle has little influence on our climate, though. To understand why, imagine putting a pot of water on the stove and then turning the burner beneath it on and off each second. If you measured the temperature of the water, you would not see it changing much each second. The thermal inertia of the water keeps the temperature from varying quickly, so it does not respond to rapid changes in heating. In our climate system, the thermal inertial of the ocean prevents the climate from significantly responding to eleven-year solar variations, just as it prevents the climate from instantly warming up in response to increased radiative forcing (this was discussed in Section 6.1).

In order for the Sun to be responsible for the recent warming, there would need to be a sustained, long-term increase in the solar constant over the past few decades. The measurements show no evidence of this. Another reason to discount the Sun as an explanation is that an increase in solar output would warm the entire atmosphere.

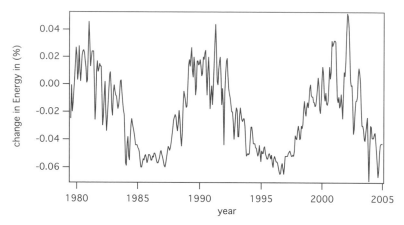

**Figure 7.1**   Percentage change in monthly values of energy in ($E_{in}$) due to changes in the brightness of the Sun, based on the analysis of Frölich and Lean. Seasonal changes in the Earth-Sun distance have been removed (adapted from Forster et al., 2007, figure 2.16).

This is not happening. Rather, measurements from weather balloons and satellites show that the stratosphere (the region of the atmosphere beginning at an altitude of 10 km or so) has cooled over the past few decades. Thus, we can conclude with high confidence that the rapid warming of the past few decades is not caused by a brightening of the Sun.

The Sun's influence on climate prior to the middle of the twentieth century is more difficult to determine because there were no satellite measurements of the solar constant. Instead, the Sun's output for this period must be inferred indirectly from other measurements, such as the number of sunspots, which people have counted for many hundreds of years, or from chemical proxies such as the carbon-14 content of plant material. The most recent analyses of these records suggest that the Sun has brightened over the past few hundred years, and this can potentially explain some of the gradual warming of the eighteenth, nineteenth, and early twentieth centuries. As I discussed in Chapter 6, this has led to a positive radiative forcing with an estimated magnitude of $+0.05$ W/m$^2$, which is small compared to net radiative forcing from human activities ($+2.3$ W/m$^2$).

## 7.3 The Earth's orbit

The solar constant is determined not just by the energy emitted by the Sun but also by the Earth-Sun distance. If, for example, the Earth moved closer to the Sun, then the solar constant would increase even if the brightness of the Sun did not change.

This is relevant because the Earth's orbit is not a perfect circle: It is an ellipse whose *eccentricity* – the ratio of the length of the ellipse to the width – varies with time. Over the course of 100,000 years or so, the orbit cycles between an orbit that is slightly more eccentric (more elliptical) and one that is less eccentric (more circular)

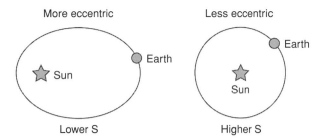

**Figure 7.2**  Schematic illustrating how the eccentricity of the Earth's orbit (how elliptical it is) varies with a period of 100,000 years or so.

(Figure 7.2).[1] As the orbit becomes more elliptical, the average Earth-Sun distance increases and the average amount of solar energy falling on the Earth decreases. For the Earth's orbit, this causes the annual average solar constant to vary by approximately 0.5 W/m$^2$ over the 100,000-year cycle.

Other aspects of the Earth's orbit can also vary, such as the timing of the closest approach of the Earth to the Sun (also known as the *perihelion*). Today, the Earth is closest to the Sun during January, when it is wintertime in the northern hemisphere. Over the next 23,000 years, the date of closest approach will cycle through the entire year. In roughly 11,500 years, the Earth will be closest during July, and in 23,000 years it will again be January.

Another important variation is the tilt of the Earth (also known as the *obliquity*). Today, the Earth's spin axis is tilted 23.5° from vertical (Figure 7.3). However, over the next 41,000 years, the Earth's tilt will complete a cycle through a range of tilt angles from 22.3° to 24.5°.

Changing the date of closest approach to the Sun or the tilt of the Earth does not change the Earth-Sun distance, so it does not change the solar constant. Rather, these changes change how sunlight is distributed over the planet, in both latitude and season. For example, increasing the tilt of the planet increases the amount of sunlight hitting the polar regions and decreases the amount hitting the tropics. Such changes can alter the climate.

We see in the paleoclimate record nearly perfect agreement between the ice-age cycles (Figure 2.13) and the variations in the Earth's orbit. As I discussed earlier in this chapter, the growth of big, continental-scale ice sheets, such as existed during the last ice age, is determined by high-latitude summertime temperatures – because this determines whether snow that falls during the winter survives the subsequent summer. Orbital variations regulate how sunlight is distributed over the planet and over the seasons, so they play a key role in regulating these temperatures. These orbital variations and the climate effects that follow are often referred to as *Milankovitch cycles,* after Serbian mathematician Milutin Milankovitch, who was the first one to recognize that the ice-age cycles corresponded to variations in the Earth's orbit.

---

[1]  When we specify the length of a cycle (e.g., the 100,000-year eccentricity cycle), this is the length for the eccentricity to execute one complete cycle. This means that it takes 50,000 years for the eccentricity to vary from its maximum value to its minimum, and another 50,000 years to return to its starting value.

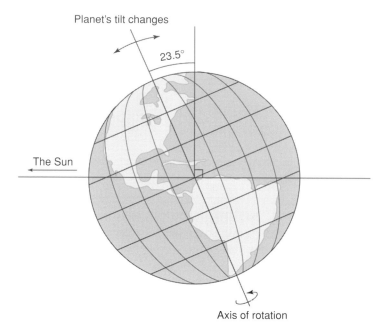

Planet's tilt changes

23.5°

The Sun

Axis of rotation

**Figure 7.3**  Schematic illustrating how the obliquity of the Earth (the tilt of the spin axis away from a line perpendicular to the orbital plane, the plane defined by the Earth's orbit) varies with a period of 41,000 years.

But while these orbital variations are critical in ice ages, are they responsible for the warming of the past few decades? They are almost certainly not. These orbital variations are so slow that it takes thousands or tens of thousands of years to make any significant change in the amount of or distribution of incoming sunlight. The warming of the past century has been much too fast to be caused by these slow orbital variations. The warming must be due to other causes.

# 7.4 Internal variability

Changes in the output of the Sun or in the Earth's orbit are examples of *forced variability*: changes in the Earth's climate in response to a radiative forcing. However, the Earth's climate system is so complex that it can also change without any external factors driving it. Such changes, which are caused by the internal physics of the system rather than external changes in the planet's energy balance, are often referred to as *internal variability*.

A good example of internal variability that you can put on your desk is the "drinking bird," a toy in which a toy bird oscillates between standing straight up and rotating over and sticking its beak into a glass of water.[2] The drinking bird relies on internal physics to drive this oscillation; its periodic behavior is not externally forced.

The best-known example of internal variability in our climate is the El Niño/Southern Oscillation (ENSO). El Niño events, which make up the warm phase of

---

[2]  See www.andrewdessler.com/chapter7 for links to videos of the drinking bird.

ENSO, occur every few years and last a year or so. During these events, the Earth warms several tenths of a degree Celsius. El Niño's opposite is La Niña, and during those events the Earth cools several tenths of a degree. These ENSO events cause a temporary temperature change every few years but no long-term changes in the climate. Figure 2.4, for example, shows the dramatic warming that the Earth experienced during the El Niño of 1998 as well as the cooling experienced during the La Niña of 2008. In fact, many of the short-term variations in the temperature record can be traced back to ENSO events.

ENSO is the dominant and best-known source of internal variability in the climate system. However, other modes of variability, with names like the Pacific Decadal Oscillation (PDO) or the Atlantic Multi-decadal Oscillation (AMO), are thought to exist, although they are less well understood. So the relevant question is this: Could the warming of the past few decades be due to one of these modes of internal variability? We can rule out ENSO as the cause because that cycle lasts a few years at most, so it could not cause a warming trend lasting several decades.

The proxy data on climate variation before the past two centuries can help answer whether another mode of internal variability – one that lasts for decades, not years – may be responsible. Human activities probably had minimal impact on climate before 1800, so climate paleoproxy data before that time should provide a good picture of recent natural climate variability. As we could see in Figure 2.15, the record between 1000 AD and 1800 AD shows nothing similar to the rate and magnitude of warming of the twentieth century. Thus, the paleoproxy data do not support internal variability as a cause of the recent warming.

We can also gain insight into natural climate variability by running climate models without any human greenhouse-gas emissions. In these simulations, climate models exhibit variations in global average temperature from year to year and decade to decade that are similar to those seen in the climate proxy data before about 1800, but they produce nothing resembling the rapid warming of the past century. The final argument against internal variability is that no one has identified any physical mechanism that would explain the warming.

Ultimately, we cannot exclude internal variability the same way we can definitively exclude, say, a brightening Sun. However, there is basically no evidence supporting this explanation, either. So internal variability is like a suspect in a criminal investigation who has no alibi but for whom there is also no evidence linking him to the crime. You cannot exclude the suspect, but you would be hard-pressed to convict him based only on a lack of an alibi.

## 7.5 Greenhouse gases

The last potential explanation for the recent warming is the increase in greenhouse gases in our atmosphere. Chapter 4 gave the physical explanation for why an increase in greenhouse gases would be expected to warm the planet, and Chapter 5 discussed in detail why this increase is almost entirely due to human activities.

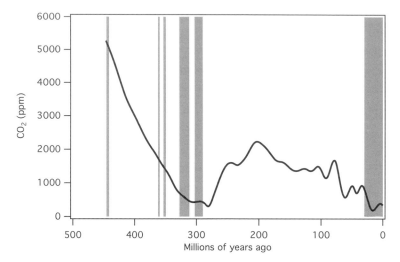

Figure 7.4 Atmospheric carbon dioxide over the past few hundred million years. The gray bars indicate times when ice existed on the planet (based on Royer, 2006, figures 1 and 2).

This physics is neither new nor complex; it was first recognized in the early nineteenth century, and our knowledge of it has been refined by nearly two centuries of work by thousands of scientists, including luminaries such as Fourier, Tyndall, and Arrhenius. In agreement with these simple physical arguments, the geologic record over the past 500 million years shows a strong correlation between temperature and atmospheric carbon dioxide. An example of this can be found in Figure 7.4, which shows periods of widespread glaciation (indicated by gray bars in the figure) when carbon dioxide was low and less ice when carbon dioxide was high.[3] Since ice extent is a proxy for temperature, we can conclude that variations in carbon dioxide and temperature have generally been associated with each other for much of the Earth's history.

## An aside: How does science deal with outliers?

Figure 7.4 also provides a good example of why climate science is hard. Although most glaciations are associated with low carbon dioxide, the eagle-eyed among you will notice that approximately 450 million years ago there was a glaciation when carbon dioxide levels were greater than 5,000 ppm. Such a point is known as an *outlier* – a point that does not agree with the rest of the data. So-called climate skeptics might take this single point and argue that it disproves the connection between climate change and greenhouse gases. Is that a reasonable conclusion? What would a scientist think about this outlier?

There are several possible explanations for the outlier. First, the theory connecting carbon dioxide with climate may indeed be wrong, as the skeptic suggests. Second,

---

[3]  Over this same time, the Sun brightened by 6 percent or so. Thus, the climate associated with a certain level of carbon dioxide a few hundred million years ago would be cooler than the climate would be for the same amount of carbon dioxide today. This explains why glaciations were occurring hundreds of millions of years ago with carbon dioxide abundances higher than today's.

the data may be wrong – perhaps there was no glaciation, or maybe carbon dioxide was really much lower than suggested by the proxy data. After all, we are trying to infer the conditions on the planet nearly half a billion years ago, and there are lots of ways that the proxy data could mislead us. Finally, both the data and greenhouse gas theory could be right: there may have been something else offsetting the warming from carbon dioxide. For example, massive volcanism could have injected enough aerosols into the atmosphere to lead to low temperatures despite high abundances of carbon dioxide.

In his seminal work, *The Structure of Scientific Revolutions*, Thomas Kuhn described how incorrect scientific theories accumulate anomalies – places where the observations do not match theory. These anomalies accumulate until there are so many that the theory is simply no longer tenable, and a scientific revolution overthrows the old theory and replaces it with a new one.

For example, at the beginning of the twentieth century, physics was in trouble. The classical theory of physics could not explain several well-validated observations, including the $T^4$ dependence of blackbody radiation (discussed in Chapter 3), atomic and molecular spectra, and the photoelectric effect. Eventually, it became apparent that classical physics simply did not work at the atomic level, and a scientific revolution occurred. What emerged was a new paradigm, in which quantum mechanics ruled small, atomic domains, and classical physics ruled our macroscopic, everyday world.

It is important to recognize that outliers occur in all fields, not just climate science. For example, you can find – contrary to expectations – people who smoke four packs of cigarettes each day yet who live to be ninety years old. Such anomalies frequently allow scientists to refine and extend their theories: Given that smoking causes cancer, why are some people less susceptible than others? Is it just luck, or is there a physiological basis? Importantly, though, the existence of some smoking anomalies does not cause scientists to reject the underlying idea that smoking is bad for your health.

The question that each scientist must ask individually, and the scientific community must ask collectively, is whether a particular theory has accumulated enough anomalies that it is no longer tenable. At present, there are not enough anomalies like the glaciation 450 million years ago to reject the dominant theory that greenhouse gases play a major role in determining our climate. But scientists are always looking for new anomalies – and if enough of them accumulate, eventually this theory of climate will be replaced by another one.

From a practical standpoint, though, no one expects that to occur. The theory that carbon dioxide exerts a strong influence on climate is so successful in predicting so many aspects of our climate that it is quite unlikely that the theory will turn out to be substantially wrong. This is akin to our views on smoking and lung cancer. Although it is possible that future research may disprove the link, it is a very, very unlikely eventuality.

---

Another example of the correlation between greenhouse gases and temperature is an event roughly 55 million years ago known as the Paleocene-Eocene Thermal Maximum or PETM, which was discussed in Section 2.2.2. This event began with a massive release of either carbon dioxide or methane, which in turn led to an increase in the Earth's global average temperature of 5–9 °C over the following few thousand

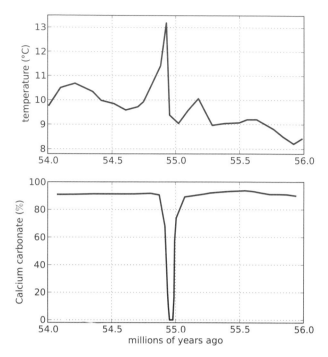

**Figure 7.5** Temperature during the Paleocene-Eocene Thermal Maximum as a function of time from various sites (top panel); calcium carbonate content of ocean sediments (bottom panel) (adapted from Jansen et al., 2007, figure 6.2).

years (top panel of Figure 7.5). The mass of carbon was so immense that when it dissolved into the oceans, the oceans became significantly more acidic (as I discussed in Chapter 5, carbon dioxide forms carbonic acid after it dissolves in water). This in turn dissolved calcium carbonate (the material that makes up shells) in the sediments at the bottom of the ocean (bottom panel of Figure 7.5).

The temperatures remained elevated for 100,000 years or so, which is about the length of time it takes the carbon cycle to fully remove the carbon from the atmosphere. Interestingly, the mass of carbon released, a few thousand gigatons, is comparable with the amount contained in all of the Earth's fossil fuels. Thus, the PETM is frequently viewed as a good analog to what will happen if humans burn all of the fossil fuels over the next few centuries. An important difference, however, is that humans are on a pace to release the carbon over several hundred years, whereas it was released during the PETM over several thousand years. Thus, we can expect even more rapid warming over the next few centuries than that experienced during the PETM, which was a period of – geologically speaking – very rapid warming.

The association between carbon dioxide and temperature is even clearer over the past few hundred thousand years. Figure 2.11 shows how carbon dioxide and temperature varied in lock step as the Earth cycled between ice ages and warm interglacials. The association between temperature and carbon dioxide, however, is a bit more complicated than the plot may at first suggest. There is strong evidence that ice-age cycles are initiated by small variations in the Earth's orbit (as discussed in Section 7.3). However, the changes in sunlight falling on the Earth in response to these slight orbital changes are too small to explain the wide temperature swings during ice-age cycles. Something must be helping the orbital variations produce the observed variations.

What is missing is carbon dioxide. The small initial warming from orbital variations leads to increased levels of carbon dioxide through mechanisms that have not yet been unambiguously identified. The increase in carbon dioxide leads to further warming. In other words, the orbital variations are the forcing, and carbon dioxide is acting here as a feedback that amplifies the small initial warming from the forcing.

We are not exactly sure what process releases carbon to the atmosphere as the climate warms during ice-age cycles. As we learned in Chapter 5, the two biggest (non-human) sources of carbon for the atmosphere are the land biosphere and ocean. For both of these reservoirs, there are plausible mechanisms that could explain why warmer temperatures would release carbon to the atmosphere, but our confidence in the details of the carbon cycle's response to climate change is low.

### An aside: A skeptical argument

During the ice ages, carbon dioxide began rising after the temperature. This proves that carbon dioxide responds to temperature, and not the other way around. Ergo, carbon dioxide cannot be causing the present-day warming.

The problem with this argument is that it misunderstands the difference between a climate forcing and a feedback. When you have a forcing, such as the Sun getting brighter or the addition of greenhouse gases to the atmosphere, the temperature change occurs after the forcing is applied to the climate system.

A feedback, however, is more complicated. There is an initial temperature change, followed by the feedback mechanism, followed by additional warming. For the ice-albedo feedback diagramed in Figure 6.8, there is an initial warming, followed by a loss of ice, followed by increased absorption of solar energy, followed by more warming. In this case, warming is first, then the ice melts. In this case, it is wrong to conclude that, because the temperature change occurs first, the melting ice has no effect on the climate. It does.

The key point here is that carbon dioxide is presently a climate forcing – humans are adding it to the atmosphere, and that is causing warming. During the ice-age cycles, however, carbon dioxide acted as a feedback – a warming planet led to the release of carbon dioxide, which then caused additional warming.

A final link between greenhouse gases and climate comes from climate models. Simulations of the twentieth century by climate models that exclude the observed increase in greenhouse gases fail to simulate the increase in temperature over the second half of the twentieth century. This can be seen in Figure 7.6: the model run in Figure 7.6a includes natural forcings – primarily changes in the solar constant and volcanoes – but no human impact on climate. This calculation reproduces many of the bumps and wiggles in the record, showing that these are not due to human activity. But this simulation completely fails to capture the rapid warming that began around 1960.

The model run in Figure 7.6b includes both natural effects as well as the effects of human activities – mainly greenhouse-gas emissions but also increases in aerosols

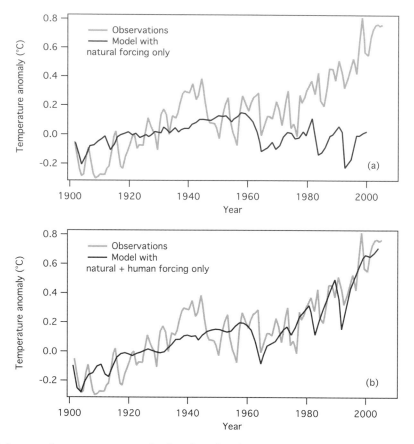

**Figure 7.6** Global mean surface temperature anomalies from the surface thermometer record (lighter curves), compared with a coupled ocean-atmosphere climate model (darker curves). The model includes (a) only nonhuman natural climate forcing, in particular solar and volcanic effects, and (b) natural forcing and human greenhouse-gas emissions, aerosols, and ozone depletion. Anomalies are measured relative to the 1901–1950 mean (the source is figure 3.12 of Dessler and Parson, 2010, which is an adaptation of figure TS.23 of Solomon et al., 2007).

and decreases in stratospheric ozone. This model captures the rapid warming since 1960 that the model with only natural forcing fails to simulate. This suggests that human greenhouse-gas emissions, volcanic effects, and solar effects have all contributed to global temperature changes of the past century but that greenhouse-gas emissions are responsible for most of the rapid late-twentieth-century warming.

## 7.6 Putting it all together

As we learned in Chapter 2, the Earth's climate has varied more or less continuously for the past several hundred million years, and probably for the entire history of the planet. Obviously, most of these variations have nothing to do with human activities.

Thus, when we consider the recent warming, the first thing we must do is investigate whether today's warming is due to natural variations. In so doing, most natural explanations can be decisively eliminated (e.g., continental drift, orbital variations, variations in the output of the Sun). Internal variability cannot be eliminated, but there is little evidence to support that as an explanation.

In contrast, there is overwhelming evidence that the increase in greenhouse gases is the cause of the recent warming. There is strong theoretical evidence that greenhouse gases warm the planet, including the simple arguments detailed in Chapter 4 and the more sophisticated calculations of climate models. There is also observational evidence that carbon dioxide has played a key role in our climate over the past 500 million years.

Taken together with the lack of a competing hypothesis, the totality of evidence that carbon dioxide is the main cause of the recent warming makes a compelling case. Reflecting this, the 2013 report of the Intergovernmental Panel on Climate Change came to the following conclusion:

> It is extremely likely that more than half of the observed increase in global average surface temperature from 1951 to 2010 was caused by the anthropogenic increase in greenhouse gas concentrations and other anthropogenic forcings together.

Note that this statement is carefully caveated in three ways. The first is the phrase *more than half.* This makes it clear that greenhouse gases are not the only factor that can influence the climate. As we have explored in this chapter, there are other factors that can influence the climate, and some of these (e.g., solar variations, internal variability) may have been minor contributors to the recent warming. However, increases in greenhouse gases are responsible for the majority of the observed warming.

The second caveat is the specified time period, 1951–2010. It is only during this period that our observations are sufficient to rule out alternative explanations for the warming. It is certainly possible that greenhouse gases were the dominant cause of the warming prior to this period, but that cannot be proven to the high standards required by the scientific community. This leaves the pre-1951 warming to be, officially at least, unattributed.

The third caveat concerns the words "extremely likely." The Intergovernmental Panel on Climate Change uses a set of carefully defined terms to express confidence. In the parlance of the IPCC, *extremely likely* denotes a confidence of 95 percent. This acknowledges that the mainstream scientific view may indeed be wrong (unlike the evidence that the globe is warming, which the IPCC describes as *unequivocal*). However, the chance is small – about one in twenty.

---

### An aside: Evolution of the IPCC's statements on warming

The IPCC has produced five major assessments of the science of climate change since 1990 – one about every six years. Below you can see the evolution of the IPCC's key statement on the attribution of the observed warming to humans – from a relatively weak statement in 1990 to a strong attribution in 2013.

1990: The size of the observed warming "is broadly consistent with predictions of climate models, but it is also of the same magnitude as natural climate variability.

Thus the observed increase could be largely due to this natural variability." This statement reflects the fact that climate science was in its infancy at that time. Satellite measurements had been available for little more than a decade, computers were slow, and there were few climate scientists in the world working on the problem. As a result, it was not possible at that time to demonstrate a clear human impact on the climate.

1995: "The balance of evidence suggests a discernible human influence on global climate."[4] In the time since the 1990 report, many advances had occurred in climate science. Most of the new evidence suggested that humans were having some effect on the climate, although an argument could still be made that the warming was mostly natural. The statement did not include any quantification of the human influence.

2001: "Most of the observed warming over the last 50 years is likely to have been due to the increase in greenhouse gas concentrations." This statement makes a much more definitive statement about the role of humans than the earlier statements did. This reflected improvements in our observations of the climate system, improvements in computers and climate models, and advances in our theoretical understanding of the planet. The word *likely* denotes a 66 percent chance (two out of three) that the statement is true.

2007: "Most of the observed increase in global average temperatures since the mid-twentieth century is very likely due to the observed increase in anthropogenic greenhouse gas concentrations." The 2007 statement is essentially identical to the 2001 statement, except it projects a higher level of confidence by using the words *very likely,* which denotes a nine out of ten chance that the statement is true.

2013: "It is extremely likely that more than half of the observed increase in global average surface temperature from 1951 to 2010 was caused by the anthropogenic increase in greenhouse gas concentrations and other anthropogenic forcings together." The 2013 statement continues the growth in our confidence in the statement: *likely* in 2001 to *very likely* in 2007 to *extremely likely* in 2013.

What is most striking, to me at least, is how climate science has actually not changed much over the years and decades. What Arrhenius thought in 1896, what scientists studying the climate thought in the 1950s, 1960s, and 1970s, and what the IPCC report described in 1990 is basically what we think in 2013. The only difference is that our confidence in this understanding has vastly improved.

The overall stability of climate science should provide us with great confidence in it. This is because important scientific ideas are constantly retested by scientists, so the longer an idea survives, the more likely it is to be correct. And as a perusal of the IPCC reports shows, most of the major claims of climate science have indeed survived a long time.

Any analysis of the cause of the recent warming should also be clear about what is *not* evidence that greenhouse gases are the primary cause of the recent warming. The case for greenhouse gases is not built on the argument that the present temperature of the Earth is exceptional. In fact, we know that the Earth has been much warmer

---

[4] The evolution of this statement is documented in a great article by Houghton (2008).

than it is today. As Figure 7.4 shows, over much of the past 500 million years the Earth was so warm that there was no ice anywhere on the planet. Nor is the case for greenhouse gases built on the argument that the rate of today's warming is exceptional. It may be that today's warming is indeed without precedent, but we simply do not have the data covering the entire Earth's history to prove that. Nor is the case built on the occurrence of extreme events, such as the extreme Atlantic hurricane season of 2005, Superstorm Sandy in 2012, or the 2003 European heat wave. Rather, the case for greenhouse gases is built on a thorough examination of all of the possible explanations. The fact that the increase in greenhouse gases can explain the warming, combined with the distinct lack of any legitimate competing theory, leads scientists to conclude that it is extremely likely that that greenhouse gases are indeed responsible.

## 7.7 Chapter summary

- To determine a cause for the present-day warming, we examine all of the natural processes that are capable of changing our climate. Among these are continental drift, variations in the Sun, and orbital variations, which can all be decisively rejected as explanations for the present-day warming. Internal variability, such as El Niño cycles, cannot be definitively eliminated as a significant cause of long-term warming, but there is also no evidence that it is.
- There is abundant evidence that the increase in greenhouse gases, which is due primarily to human activities, can explain the present-day warming. There is strong theoretical evidence that greenhouse gases warm the planet, including the simple arguments detailed in Chapter 4. And sophisticated calculations by climate models are only capable of reproducing the warming of the past half-century if the increase in greenhouse gases is included. There is also strong observational evidence that carbon dioxide has played a key role in our climate over the past 500 million years.
- On the basis of this evidence, the IPCC concluded in its 2013 report that "It is extremely likely that more than half of the observed increase in global average surface temperature from 1951 to 2010 was caused by the anthropogenic increase in greenhouse gas concentrations and other anthropogenic forcings together."

## Terms

Continental drift
Eccentricity
Forced variability
Internal variability
Milankovitch cycles
Obliquity
Outlier
Perihelion

# Additional reading

The Working Group I report of the IPCC's Fifth Assessment describes, at varying levels of detail, the evidence attributing the recent warming to humans. For the most detail, see Chapter 10 of the report. For a less detailed overview, see Section TS.3 of the Technical Summary. And, for a short, high-level discussion, see Section D.3 of the IPCC's Summary for Policymakers (you can download all of these at www.ipcc.ch/report/ar5/wg1/).

SkepticalScience.com has several useful write-ups that summarize the evidence that humans are responsible for most of the recent warming (www.skepticalscience .com/its-not-us-basic.htm, www.skepticalscience.com/empirical-evidence-for-global-warming.htm).

See www.andrewdessler.com/chapter7 for additional resources for this chapter.

# Problems

1. a) List all of the physical processes that can alter the climate.
   b) For all processes in part (a) except greenhouse gases, explain why they are unlikely to be the cause of the warming over the past few decades.
   c) List the evidence that greenhouse gases are responsible for the recent warming.
2. What did the IPCC say in its 2013 report about whether humans are causing climate change? What are the three caveats in the statement?
3. Why are feedbacks (e.g., increases in water vapor) not discussed as potential causes of climate change?
4. Explain the physical mechanism for the occurrence of ice ages. Make sure you explain the role of carbon dioxide and its timing with respect to the temperature change.
5. Critique this statement: "It is clear that it was warmer around 1000 AD, during the Medieval Warm Period, than it is today. Therefore, humans cannot be causing today's warming." Assume that the claim that the Medieval Warm Period was warmer than today is correct (it may have been, but it is debatable). Is this argument correct? Why or why not?
6. What are the three ways that the Earth's orbit varies? How does each variation affect the climate?
7. Explain how the Paleocene-Eocene Thermal Maximum provides support for the claim that today's warming is caused by humans.
8. How does continental drift affect our climate?

# Predictions of future climate change

In Chapter 6, we discussed the concept of radiative forcing, which is an imposed change in planetary energy balance. In response, the planet's temperature adjusts so as to restore energy balance. Thus, if we can predict how radiative forcing will evolve in the future, we can then estimate how much climate change we will experience.

Predicting future radiative forcing basically comes down to predicting how much greenhouse gas and aerosols will be emitted into the atmosphere each year from human activities. Such projections, known as *emissions scenarios,* therefore form the backbone of our predictions of climate change. In this chapter, I describe how they are constructed and what they tell us about our future climate.

## 8.1 The factors that control emissions

At its simplest, the amount of greenhouse gas released by a society is determined by the total amount of goods and services consumed by that society. This is true because the production of any good or service – be it a car, an iPhone, a university lecture, a cheeseburger, or an hour of tax consulting – requires energy. And energy is mostly derived from the combustion of fossil fuels, which leads to the release of carbon dioxide to our atmosphere. The emissions of other greenhouse gases and aerosols also generally scale with the amount of consumption, although the causal linkages may not be as direct.

The total value of goods and services produced by an economy is known as the *gross domestic product,* abbreviated GDP. Thus, total emissions by a society are basically set by that society's GDP. If the GDP doubles, then we expect emissions to double, as long as everything else remains the same. This strong link between GDP and emissions can be seen during recessions. For example, during the severe economic downturn of the late 2000s, global carbon emissions posted their biggest drop in more than forty years as the global recession froze economic activity and slashed energy use around the world.

Rather than consider GDP as a whole, it is useful to break it into the product of two factors: population and affluence. It should be obvious that GDP scales with population. Every person in a society consumes goods and services, so if the population doubles (and everything else remains the same), then total GDP will also double. Emissions will therefore also scale with population – so emissions double if the population doubles.

In addition to the number of people, how rich each person is also matters because, as people get richer, they consume more. To illustrate the affect of affluence on GDP and emissions, consider the following three families. The first is a family of four who live as subsistence farmers in sub-Saharan Africa. This family lives in a small one-room house without electricity or running water. They do not own a car and are too poor to buy anything but the bare necessities of life. They farm by hand or with a draft animal. Because the members of this family are so poor and consume so little, they are responsible for little greenhouse-gas emissions.

Now consider a family of four near the bottom of the economic spectrum in the United States. They live in an apartment and they own one car. Their apartment is not air-conditioned; they own a television and one or two heavy-duty electrical appliances, such as an oven. Compared with the subsistence farming family in Africa, this family is far richer and consumes far more and is therefore responsible for more greenhouse-gas emissions.

Finally, consider an upper-class family of four in the United States. This family lives in a 4,000-ft$^2$ single-family house and owns three cars (for the husband, wife, and a teenage child). The house has televisions in almost every room, several computers, VCRs, game consoles, and a rich assortment of electrical appliances. The family flies to several vacation locations every year. Because of the significant consumption allowed by their affluence, this family is responsible for more emissions than the poorer U.S. family and many, many times the emissions of the subsistence farming family.

This wealth effect leads to enormous disparities in emissions per person. In the United States, emissions are about 5 tons of carbon per person. Emissions in China are 1.7 tons per person – about one-third of U.S. per capita emissions. However, China's population is so large that they nevertheless lead the world in total carbon emissions. Emissions in Nigeria are 0.1 tons per person – about one-fiftieth of the United States – reflecting the country's poverty.

We need a third factor to convert a level of total consumption, expressed in dollars, to greenhouse-gas emissions. This last factor relates how much greenhouse gas is emitted for every dollar of consumption; it is known as the *greenhouse-gas intensity*. Putting these all together, we can now relate emissions to the factors that control it in a simple equation:

$$I = P \times A \times T \tag{8.1}$$

Here $I$ represents the total emissions of greenhouse gases into the atmosphere (these emissions then cause climate impacts, which is why emissions are represented by the letter I); $P$ is the population, $A$ stands for affluence, and $T$ stands for greenhouse-gas intensity. Affluence $A$ is GDP per person – the average amount of goods and services each person consumes – so the product of $P$ and $A$ is the GDP. The decomposition of emissions into these factors is often referred to as the *IPAT relation* or the Kaya Identity.

The greenhouse-gas-intensity term $T$ can be usefully broken down as the product of two terms:

$$T = \text{EI} \times \text{CI} \tag{8.2}$$

EI stands for *energy intensity* – the number of joules of energy it takes to generate one dollar of goods and services. The EI of an economy is primarily determined by two factors. First is the mix of economic activities that make up the economy. For example, it takes much more energy for a steel mill to produce one dollar's worth of steel than for a university to produce one dollar's worth of teaching. The steel mill must run blast furnaces and other heavy equipment, whereas the university only requires lighting, air-conditioning, computers, and the like. More generally, industrial manufacturing has a higher energy intensity than white-collar service-oriented activities. The more industrial manufacturing an economy has, the higher its energy intensity.

The second factor in determining the energy intensity of an economy is the efficiency with which the economy uses energy. For any economic activity, there are usually several technologies to accomplish it. For lighting, for example, there is the standard incandescent light bulb (the kind with the filament) or the compact fluorescent light bulb. As described in Chapter 3, incandescent light bulbs are dreadfully inefficient, requiring 60 W of power to produce the same light as a compact fluorescent light bulb drawing 14 W. Both light-bulb technologies can light a room, but they consume vastly different amounts of energy doing it. The trade-off is that better technology is often more expensive. As a result, it takes a certain level of wealth in order to adopt the most energy-efficient technology, and the efficiency with which different countries utilize energy can vary greatly.

CI in Equation 8.2 stands for *carbon intensity* – the amount of greenhouse gas emitted per joule of energy generated – which reflects the mix of technologies used to generate energy. Put another way, it is determined by whether the economy uses coal, oil, gas, nuclear, wind, solar, etc. to generate energy. Among fossil fuels, combustion of natural gas (methane or $CH_4$) produces the least carbon dioxide per joule of energy generated. Thus, it has the lowest carbon intensity, which is one of the reasons it is often considered to be the "greenest" of the fossil fuels. Oil produces more carbon dioxide per joule than methane, so it has a higher carbon intensity. The most carbon-intensive fossil fuel is coal – it produces roughly twice the carbon dioxide per joule as methane – which explains why many people who are concerned with our climate are opposed to the construction of new coal-fired power plants. Energy sources also exist that produce no carbon dioxide, such as hydroelectric, nuclear, wind, and solar energy sources.

For a country such as France, which generates most of its electricity from nuclear energy, the carbon intensity will be smaller than for a country such as China or the United States, which both rely heavily on coal for electricity.

---

## An aside: Check the units!

One of the most powerful ways to check your work is to make sure that the units in a problem work out. We do this now to close the loop on our understanding of the factors that regulate carbon emissions.

Population is obviously the number of people. Affluence is dollars of GDP per person. The product of population and affluence is therefore GDP, which has units of dollars:

$$\text{\# of people} \times \frac{\$\text{GDP}}{\text{person}} = \$\text{GDP}$$

Energy intensity has units of joules per dollar, and carbon intensity has units of carbon dioxide emitted per joule. Greenhouse-gas intensity is the product of energy

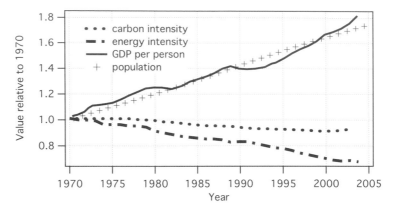

Figure 8.1 Population, affluence, carbon intensity, and energy intensity for the entire world, relative to values in 1970 (adapted from IPCC, 2007b, figure 2).

intensity and carbon intensity, and therefore it has units of carbon dioxide emitted per dollar:

$$\frac{J}{\$GDP} \times \frac{CO_2}{J} = \frac{CO_2}{\$GDP}$$

Finally, the product of population, affluence, and technology has units of carbon dioxide emitted:

$$\text{\# of people} \times \frac{\$GDP}{\text{person}} \times \frac{CO_2}{\$GDP} = CO_2$$

## 8.2 How these factors have changed in the recent past and how will they change in the future

In the last section, emissions of carbon dioxide were deconstructed into the controlling terms: population, affluence, energy intensity, and carbon intensity. Let us look at how these terms have changed over the past few decades and how they might change in the future.

### 8.2.1 Population

Population has been rapidly increasing for the past few centuries. It took all of human history up to 1804 for the global population to reach 1 billion. The 2-billion-people mark was reached 123 years later, in 1927, and the 3-billion-people marker was reached 33 years later, in 1960. Since then, world population has been increasing by 1 billion people every twelve to thirteen years, reaching 6 billion in 1999 and 7 billion in 2011. Figure 8.1 shows that the population has increased by 80 percent over the past few decades. Today, world population is increasing by roughly 200,000 people per day, a population growth rate of approximately 1 percent/year. Most of this growth is occurring in the developing world, where fertility rates remain high.

In estimating future population change, some of the controlling factors are well known. Affluence, for example, strongly determines how many kids a woman has, with the poorest countries having the highest fertility rates. In extremely poor societies, children can be put to work at a young age and are therefore a source of income. This is generally not the case in rich countries, where children are a net drain on family resources for many years (trust me on that). In addition, high rates of childhood death in poor countries mean that parents must have many children to ensure that some of them survive into adulthood. Improvements in health care that occur as a society gets richer, however, mean that rich parents can reasonably expect their children to survive into adulthood. The amount of education that women receive is also a factor, with fertility rates declining as women become better educated and good-paying jobs become available to them as an alternative to child rearing. Our understanding of these factors gives us some ability to predict future population.

However, some events that affect population are impossible to predict. It is impossible to predict, for example, societal changes, such as a future Pope suddenly embracing birth control, causing a fertility decline among the roughly one billion Catholics. Or the occurrence of chance events, like a nuclear war or the emergence of new diseases like AIDS, which kill millions of people.

The lowest population scenarios predict population will peak at around 9 billion in the middle of the twenty-first century, decreasing thereafter and reaching 7 billion by 2100. Higher population scenarios predict population increasing, more or less continuously, reaching 16 billion by 2100. Our best estimate is that world population will stabilize during the twenty-first century around 10 or 11 billion.

## 8.2.2 Affluence

Figure 8.1 shows that affluence, measured as GDP per person, increased by 80 percent over the past few decades of the twentieth century. My personal experience backs this up. When I was a college student in the early 1980s, I did not own a cell phone or a laptop or tablet computer, my car had handcranked windows, did not have air conditioning or antilock brakes or an airbag, and only had an AM radio. I did not own a TV or videogame console. I had, in other words, a far lower standard of living than most of today's college students.

We can also see discrete political events in the affluence data, such as the 1989 collapse of the Soviet Union and the associated political upheaval in Eastern Europe, as well as various recessions. Moreover, in the past decade, the remarkable economic growth of China (with affluence growing at 10 percent/year or so) and other large developing countries has played a key role in driving global growth of consumption.

In the future, factors such as the level of education in the population, rule of law, free trade, and access to technology will be key in determining how fast affluence grows. In general, economic growth rates are highest for countries making a transition out of poverty and into the group of rich countries of the world. For example, economic growth was fastest in the United States in the late nineteenth century, in Japan after World War II, and in China today. Growth rates are lower for large, advanced societies. Based on these factors, expert predictions are that affluence will increase over the twenty-first century at 2 to 3 percent/year for developing countries and 1 to 2 percent/year for industrialized countries.

## 8.2.3 Technology

The first part of the technology term, the energy intensity term, has decreased over the past century as our society has developed more efficient ways to use energy (Figure 8.1). Some of this increasing efficiency has been driven by market forces: Because energy costs money, a more energy-efficient piece of equipment or process will reduce costs, which consumers want. As a result, just about everything you buy today is more energy efficient than the comparable version of a few decades ago. Much of this increase in efficiency is incremental, meaning that each new generation of a particular piece of equipment uses slightly less energy than the previous version. Sometimes, however, there is a revolution in technology that greatly reduces energy consumption. A good example is the revolution in lighting technology we are now experiencing. As the world switches from incandescent bulbs to compact fluorescent bulbs and LED bulbs, the amount of energy being consumed by lighting will experience a substantial one-time drop.

Changes in the mix of goods and services produced by the world's economy has also led to decreases in energy intensity. Over the past century, the fraction of the world economy based on energy-intensive heavy industry and manufacturing has declined, while the fraction based on services has increased.

Overall, energy intensity has at times decreased as fast as 2 percent/year, but the periods of fastest decreases occurred during periods of rapid economic shifts or as responses to energy price shocks. More typical values for the twentieth century were decreases of 1 percent/year. It is likely that this rate of decrease can be sustained over the coming century.

Figure 8.1 also shows that carbon intensity, the amount of carbon dioxide released per joule of energy generated, has decreased slightly over the past few decades as the world shifts from coal to cleaner natural gas. Nevertheless, coal plants continue to be built, particularly in developing countries such as China, and this prevents more rapid decarbonisation of our economy.

Continued reductions in carbon intensity can be expected given the flood of cheap natural gas that has arisen from the development and application of new drilling techniques, in particular hydraulic fracturing (more commonly known as fracking). Increases in renewables (e.g., wind and solar) are also expected to reduce carbon intensity. Depending on government policies, however, adoption of renewables could be slow or rapid, leading to minimal or large reductions in carbon intensity. These trends are expected to continue the long-term decline in coal usage, although coal consumption is still increasing in certain places, such as China.

Overall, increases in population (P) and affluence (A) have increased faster than greenhouse-gas intensity (T) has declined, leading to an increase in emissions of 75 percent between 1970 and 2005. Whether this happens in the future depends in large part on how the world's economy evolves and what the world decides to do about climate change, as we discuss in the next section.

# 8.3 Emissions scenarios

"It's hard to make predictions – especially about the future?"[1]

Although we have a good idea of the factors that control greenhouse-gas emissions, making accurate predictions of these factors is difficult. For example, predicting future population trends requires predictions of factors such as the rate of poverty, evolution of religious and social views on birth control, the rate of education of women in high-fertility regions, available healthcare in these regions, and so on.

Because of this difficulty, it is impractical to make a single prediction of future emissions. Instead, the community of experts has developed a set of alternative *emissions scenarios*. Each scenario is an internally consistent vision of one way the world might evolve in the future, and the full set of emissions scenarios is designed to span a plausible range of alternative futures.

The scenarios most recently used by the IPCC in its predictions of future climate change are known as the *Representative Concentration Pathways*, frequently abbreviated RCP. The individual RCP scenarios are named RCPx, where x is the radiative forcing in 2100. Thus, the RCP8.5 scenario has radiative forcing of 8.5 $W/m^2$ in 2100, while the RCP2.6 scenario has radiative forcing of 2.6 $W/m^2$ in 2100.

Each RCP scenario is associated with an internally consistent set of assumptions for population, affluence, and technology. For example, because people have fewer children as they get richer, the scenarios in which the world's poor become richer feature slower population growth than the scenarios in which poverty is rampant. And the development and adoption of new technology requires high economic growth to support it – so the higher the economic growth scenarios also have more rapid adoption of new and cleaner technologies.

Given an emissions scenario, the atmospheric concentrations of carbon dioxide can be calculated by feeding those emissions into a carbon-cycle model. The carbon-cycle model calculates how much of the carbon dioxide emitted to the atmosphere is absorbed by the ocean and land reservoirs. The remainder stays in the atmosphere and increases atmospheric carbon dioxide.

Figure 8.2 shows yearly carbon dioxide emissions during the twenty-first century. The RCP8.5 scenario is the most pessimistic – it assumes that humans make essentially no effort to reduce emissions. As a result, emissions increase throughout the twenty-first century, reaching 30 GtC/year by 2100 – about triple today's emissions. Emissions level off and decrease after 2150, reaching near-zero emissions by 2250.

Figure 8.3 shows that atmospheric carbon dioxide abundances increase rapidly in response to RCP8.5's huge emissions, with mixing ratios exceeding 1,800 ppm in 2200 – this is more than six times the preindustrial abundance of 280 ppm.

The RCP6 and RCP4.5 scenarios assume that the world makes some effort to reduce emissions; as a result, they have emissions peaking in the middle of the twenty-first century. This leads to atmospheric carbon dioxide stabilizing early in the

---

[1] This statement has been attributed to various people, including Niels Bohr and Yogi Berra.

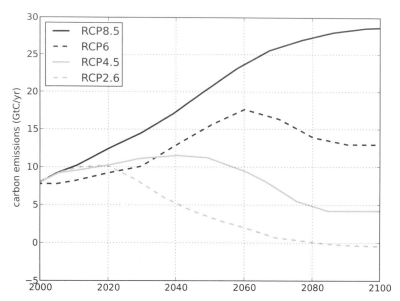

Emissions of carbon dioxide for the four emissions scenarios. The emissions are in GtC per year (adapted from figure 6 of van Vuuren et al., 2011).

next century at 750 ppmv and 540 ppm, respectively – corresponding to about three times and twice preindustrial carbon dioxide abundance.

The RCP2.6 scenario is the most optimistic scenario, with emissions peaking around 2020 and then decreasing throughout the rest of the century. After 2080, emissions actually become net negative, meaning that carbon removal from the

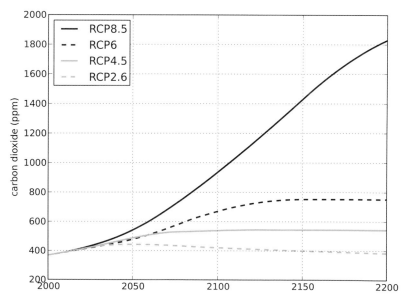

Atmospheric abundances of equivalent carbon dioxide (in ppm) for the four emissions scenarios. Equivalent $CO_2$ is the amount of $CO_2$ that gives the same radiative forcing as the full suite of radiative forcers in the atmosphere in any year (data downloaded from the RCP database: www.iiasa.ac.at/web-apps/tnt/RcpDb).

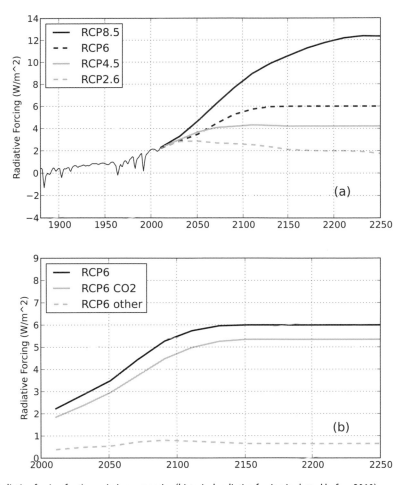

**Figure 8.4**  (a) Radiative forcing for the emissions scenarios (historical radiative forcing is plotted before 2010).
(b) Radiative forcing for the RCP6 scenario, along with the radiative forcing in that scenario just from carbon dioxide and from everything other than carbon dioxide (data downloaded from the RCP database: www.iiasa .ac.at/web-apps/tnt/RcpDb).

atmosphere (by, for example, growing plants and then burying the carbon) exceeds carbon released to the atmosphere. This leads to atmospheric carbon dioxide peaking in 2050 at 440 ppm and decreasing thereafter. By 2150 it is below present day values of 400 ppm, and by 2500, it is 327 ppm – almost back to preindustrial carbon dioxide. Achieving anything close to the RCP2.6 trajectory would require truly heroic efforts.

## 8.4 Predictions of future radiative forcing

Given the atmospheric abundances of carbon dioxide in Figure 8.3, along with abundances of other greenhouse gases and aerosols, the radiative forcing can be calculated. Figure 8.4a shows the radiative forcing predicted for each scenario over the next

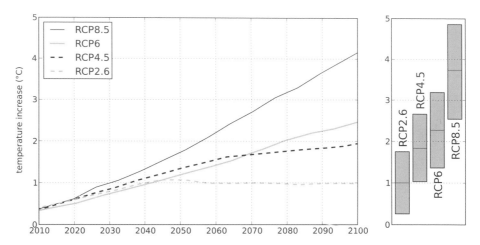

Figure 8.5 (left) Model estimates of global annual average surface temperature for the four emissions scenarios (relative to the 1986–2005 average). The bars at right are the likely temperature difference between the 1986–2005 period and the 2081–2100 period. Adapted from figure SPM.7 of IPCC [2013].

250 years. Forcing in the RCP8.5 scenario increases until about 2200, when it reaches more than 12 W/m². Forcing in the RCP6 and RCP4.5 scenarios stabilizes around 2100 at 6 and 4.5 W/m², respectively. Radiative forcing from the optimistic RCP2.6 scenario declines continuously from its peak around 2020. By 2150 in this scenario, radiative forcing has declined below present day values.

Figure 8.4b shows radiative forcing from the RCP6 scenario, along with the radative forcing broken down into the contribution from carbon dioxide and from everything else. The plot shows that, as we go into the future, carbon dioxide is responsible for virtually all of the increase in radiative forcing – the non-carbon dioxide component does not change much. The reason for this is carbon dioxide's long lifetime in the atmosphere (discussed in Chapter 5) – once emitted, carbon dioxide stays in the atmosphere for centuries. So carbon dioxide accumulates in the atmosphere like water in a stopped up sink, and the radiative forcing from it accumulates too. Other greenhouse gases, like methane, have a much shorter lifetime (methane's is about ten years), so it does not accumulate in the same way that carbon dioxide does. This explains why there is such a strong focus on carbon dioxide in policy debates over climate change.

# 8.5 Predictions of future climate

## 8.5.1 Over the next century

The estimates of atmospheric radiative forcing shown in Figure 8.4 are then input to climate models, which calculate a future climate for each scenario. These are plotted in Figure 8.5 and they show that the set of emissions trajectories translates into a wide range of future climates. The large emissions associated with the RCP8.5 scenario lead to temperature increases of 4°C over the twenty-first century, while the low emissions associated with the RCP2.6 scenario lead to temperature increases of only 1°C, mostly during the first half of the century. If you want to relate the warming to preindustrial temperatures, add about 0.8°C to each of these numbers.

Given the present political environment, the RCP2.6 scenario, which has emissions peaking around 2020, appears hopelessly optimistic. This RCP4.5 scenario, with emissions peaking around 2040 and atmospheric carbon dioxide stabilizing around 540 ppm, seems (to me, at least) the best we can hope for. That scenario yields warming of 1.8°C over the twenty-first century. Of course, we might do worse than this, and even the RCP8.5 scenario is not out of the question. Given this, a reasonable estimate of temperature increases over the twenty-first century might be 1.8–4.2°C.

It is worth noting that, despite huge differences in emissions, the scenarios predict relatively similar warming until they begin to diverge around 2040–2050. This occurs for two main reasons. First, emissions reductions require us to fundamentally rebuild our energy infrastructure. Doing this at any kind of reasonable cost means that it will take place over several decades. Thus, even the most stringent efforts to cut emissions will have only modest near-term impact (this can also be seen in emissions estimates in Figure 8.2). Second, the high heat capacity of the ocean means that any difference in radiative forcing needs to act for several decades before significant differences in surface temperatures are evident.

The upshot is that the temperature trajectory over the next few decades has already been determined largely by greenhouse-gas emissions that have already occurred, investments in energy infrastructure that we have already made, and the slow response of the climate system due to thermal inertia from the ocean.

But Figure 8.5 also clearly shows that we do have significant control over the amount of warming experienced by the end of the twenty-first century. This is one of the many aspects of climate change that make it difficult to solve. Addressing climate change will require us make investments now and in the next few decades in renewable energy and other energy efficiency technologies. But these investments will really only pay off in the second half of the century. In other words, addressing climate change requires people today to take actions that mainly benefit future generations.

The right-hand panel in Figure 8.5 shows the likely range of temperatures at the end of the twenty-first century predicted for all four RCP scenarios. This range is generated by taking the same emissions scenarios and running them through a large number of different climate models. Because each climate model handles the details of the physics of the climate system differently, the models produce slightly different results. Thus, predictions of climate at the end of the twenty-first century are uncertain because of uncertainty in which emissions pathway the world will follow and also because of uncertainties in the physics of the climate models.

An aside: Will the evolution of the climate over the twenty-first century look like the trajectories plotted in Figure 8.5?

Not really. If you look at Figure 2.2a, you will see that, over the past 130 years, the climate has generally warmed, but it also shows significant year-to-year variability, which is caused by things such as El Niño cycles and volcanic eruptions. Such variability means that temperatures can decline for a few years, even as the climate is experiencing a long-term warming (this was shown in Figure 2.5). Individual model simulations also show this year-to-year variability. The model lines plotted in Figure 8.5, however, are not the result of individual model runs. Instead, each line is the average of many model runs. In the individual model runs going into the average, the highs and lows caused by the short-term variability do not occur at the same time, so when you average many model runs together, the short-term ups and downs tend to cancel out and you get a smooth increase in temperature throughout the century. In reality, short-term variability is going to be important and we can expect the same kinds of ups and downs seen in the past 130 years to continue to occur in the future.

## 8.5.2 Climate change beyond 2100

Even though Figure 8.5 stops in Year 2100, climate change does not stop at that date. Exactly how long emissions can continue is a fiercely debated point. Fossil fuels must eventually run out, and emissions from their combustion will then cease. The range of total possible emissions until that occurs extends from lower values of 1,500 GtC to more worrying estimates of 5,000 GtC. These estimates are all well above the 370 GtC that humans have already emitted into the atmosphere over the past few centuries, and even the lowest estimates of carbon reserves would, if we burned it all, lead to a manyfold increase in atmospheric carbon. It seems likely to me that greenhouse-gas emissions will eventually cease because of concern about climate change or because technological developments make fossil fuels obsolete. But, given the deadlock in the public climate change debate, it is difficult to predict when that will occur.

In Chapter 5, we talked about the long lifetime of carbon dioxide in our atmosphere: Of carbon dioxide emitted into the atmosphere today, about 25 percent will still be in the atmosphere in several centuries, and it will take hundreds of thousands of years to remove all of the added carbon from the atmosphere (Figure 5.9). The impact of the long residence time of carbon dioxide in our atmosphere is shown in Figure 8.6a, which shows atmospheric carbon dioxide over the next 1,000 years for emissions scenarios in which atmospheric carbon dioxide increases until it reaches 550, 850, and 1,200 ppm, at which point emissions from human activity decline instantly to zero.

Even by the year 3000, eight to nine centuries after carbon dioxide emissions ceased, atmospheric carbon dioxide in all scenarios remains well above preindustrial values (280 ppm). This is simply a reflection of how long it takes for an addition to atmospheric carbon dioxide to be removed.

The long-term evolution of temperatures associated with these carbon dioxide time series are shown in Figure 8.6b. Even after emissions cease, the temperatures do not significantly decline over the next 1,000 years. This is a consequence of three factors. First, carbon dioxide remains elevated throughout the millennium, so it continues to trap heat for a very long time after emissions stop. Second, the ocean's large heat capacity means that the planet cools off very slowly because of its large heat capacity. This is the flip side of the situation in which the warming lags the carbon dioxide

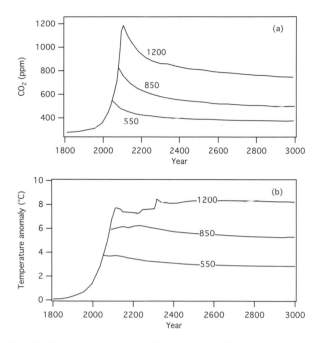

**Figure 8.6**   (a) Amount of carbon dioxide in the atmosphere as a function of time, for the next 1,000 years. Carbon dioxide emissions rise at 2 percent/year until it hits a peak abundance (550, 850, and 1,200 ppm); then emissions are decreased instantly to zero. (b) The temperature time series corresponding to each carbon dioxide time series (adapted from Solomon et al., 2009, figure 1).

increase – the cooling will lag any decrease in atmospheric carbon dioxide abundance. Third, slow feedbacks, such as the very slow destruction of the planet's big ice sheets, will act to oppose any cooling.

The important point here is that emitting large amounts of carbon dioxide to the atmosphere this century commits the planet to elevated temperatures for thousands of years. Once the temperatures rise, reducing emissions will not bring the temperature back down quickly. We can therefore think of climate change as being irreversible over any time period that we conceivably care about. This also means that actions we take today (or do not take) to curb emissions over this century will essentially determine the climate for thousands of years. It is sobering indeed to realize that people of the year 3000 or 4000 will be so affected by actions we take today.

The irreversibility of carbon dioxide emissions can be usefully contrasted with the second most important greenhouse gas, methane. Methane has an atmospheric lifetime of ten years, meaning that a few decades after emission, just about all of the methane is gone from the atmosphere. Thus, if humans ever stop emitting methane, we would be back down to its preindustrial value in a few decades.

## 8.6 Is the climate predictable?

One criticism of climate predictions goes something like this: "We cannot predict the weather next week, so why does anyone believe predictions of the climate in a hundred years?" This may sound reasonable, because it is based on the correct observation that weather predictions are only accurate a week or so into the future. However, the argument is built on a fatal flaw – it makes the mistake of equating weather predictions with climate predictions. In fact, it *is* possible to predict the climate in 100 years even if weather is only predictable for a few days.

The root cause of this conundrum is that predicting the weather and predicting the climate are fundamentally different problems. A weather forecast is a prediction of the exact state of the atmosphere at an exact time: "At 8 AM tomorrow, the temperature in Washington, DC, will be 3°C, and it will be raining." If you get the time of an event wrong – for example, you predict rain for 8 AM but it does not rain until 6 PM – then you have blown the forecast. If you predict rain for the Washington, DC, area but the rain falls 50 km to the west in Northern Virginia, then you have blown the forecast. And if your temperature is off by a few degrees and snow falls instead of rain, and it completely snarls traffic on Interstate 495, then you have *really* blown the forecast.

A climate prediction, in contrast, does not require predicting the exact state of the atmosphere at any particular time; instead, it requires predicting the *statistics* of the weather over time periods of years. Thus, a climate prediction for the month of March for the years between 2080 and 2090 for a particular location might be as follows: average monthly temperature of 12°C, with an average high of 16°C and an average low of 5°C; monthly average precipitation of 6.0 cm; and so on.

Being unable to make a prediction of the exact state of a complex system (e.g., the weather) does not preclude the ability to predict the statistics of the system (e.g., the climate). As an analogy, consider that it is virtually impossible to predict the outcome of a single flip of a coin. However, the statistics of coin flips are trivial: If you flip a coin 100 times, I can tell you that you will get approximately fifty heads and fifty tails. In other words, the inability to accurately predict any single coin flip does not preclude the ability to predict the long-term statistics of the coin.

To make this point more concretely, answer the following question: "Is it going to be hotter in Texas next January or next August?" If you know Texas weather, you can predict with 100 percent certainty that August is the hotter month, and you can make this prediction months, years, or decades in advance. Think about that for a minute: You just made a climate prediction that is valid years in advance – far beyond the ability to predict weather.

More technically, weather forecasts belong to a class of problems known as initial value problems. This means that, to make a good prediction of the future state of the system, you must know the state of the system now. If you have a marble rolling down a slope, and you want to predict where it will be in one second, you need to know where it is now to make that prediction. Similarly, to make a good weather forecast for tomorrow, you have to accurately know the state of the atmosphere today. The state of today's atmosphere is then put into a forecast model, which turns out a prediction of tomorrow's atmosphere. However, small errors in our knowledge of today's atmosphere grow exponentially, so that a forecast more than a few days in the future is dominated by the errors in our knowledge of today's atmosphere. That is why weather forecasts break down after a few days.

Climate forecasts are a class of problems known as boundary value problems. This type of problem does not require knowledge of today's atmospheric state but rather requires a knowledge of the radiative forcing of the climate. This is why, for example, we can predict with 100 percent certainty that August in Texas will be on average hotter than January in Texas. We know this because we know that more sunlight falls on Texas and the rest of the northern hemisphere during summer, leading to higher temperatures.

Increases in greenhouse gases also increase the heating of the surface, although by infrared radiation rather than visible. Thus, we can have confidence that, if we add greenhouse gas to the atmosphere, the increase in surface heating will warm the planet – just as we can predict that summer will be hotter than winter.

One should not take this to mean that predicting the climate is an easier problem than predicting the weather, only that they are different problems. Some aspects of the climate problem are, in fact, harder than the weather problem. For example, because weather forecasts cover only a few days, weather models can assume that the world's oceans and ice fields do not change. Climate models, however, cannot make this assumption, because both the world's oceans and its ice fields can significantly change over a century. Climate models must therefore predict changes in these and other factors in order to accurately predict the evolution of the climate system over a century.

# 8.7 Chapter summary

- Prediction of future climate requires predictions of future emissions of greenhouse gases from human activities. Such predictions are known as emissions scenarios.
- The factors that control emissions are population ($P$), affluence ($A$), and greenhouse-gas intensity ($T$). This is expressed by what is known as the IPAT relation: $I = P \times A \times T$, where $I$ is carbon dioxide emissions.
- Greenhouse-gas intensity is the product of energy intensity and carbon intensity. Energy intensity reflects the efficiency with which the society uses energy as well as the mix of economic activities in the society, with units of Joules of energy consumed per dollar of economic output. The carbon intensity reflects the technologies the society uses to generate energy, and it has units of carbon dioxide emitted per Joule of energy produced.
- Because predictions of the future are so uncertain, scientists have constructed a set of plausible, alternative scenarios of how the world might evolve. Taken as a group, these Representative Concentration Pathways span the likely range of future emissions trajectories.
- Putting these emissions scenarios into a climate model yields predictions of warming over the twenty-first century of 1.8 to 4.2°C. This is much larger than the warming of 0.8°C that the Earth experienced over the course of the twentieth century.
- Climate change does not stop in Year 2100. Carbon dioxide stays in the atmosphere for centuries after it is emitted, so large emissions of carbon dioxide this century will cause the Earth's temperatures to remain elevated for thousands of years.

- Even though weather is not predictable beyond a few days, we can nevertheless make climate predictions decades in advance. A climate prediction is a prediction of the statistics of the system. For many complex systems, predicting the statistics is easier than predicting the specific state of the system.

# Additional reading

Chapter 11 (Kirtman et al., 2013) and chapter 12 (Collins et al., 2013) of the IPCC's 2013 report detail, respectively, short-term and long-term predictions of climate change.

Section 12.3 of Collins et al. (2013) describes the RCP scenarios. For a more readable description, see SkepticalScience: www.skepticalscience.com/rcp.php

D. Archer, *The Long Thaw: How Humans Are Changing the Next 100,000 Years of Earth's Climate* (Princeton, NJ: Princeton University Press, 2010). Among the many things covered in this book is the very long-term evolution of climate change.

See www.andrewdessler.com/chapter8 for additional resources for this chapter.

# Terms

Carbon intensity
Emissions scenario
Energy intensity
Greenhouse-gas intensity
Gross domestic product (GDP)
IPAT relation
Representative Concentration Pathways

# Problems

1. a) Someone asks you about how much the climate will warm over the next 100 years if we do nothing to address climate change. How do you answer?
   b) If the amount of carbon dioxide and other greenhouse gases stopped increasing today and were held constant into the future, how do you think the climate would change over the next century?
2. a) Define each term in the IPAT identity.
   b) What are the units of each term? Show how the units cancel so that the $I$ term has units of emissions of greenhouse gases.
3. a) The $T$ term can be broken into two terms. What are these two terms, and what are their units?

b) If we switch from fossil fuels to solar energy, which of the terms changes, and does this term increase or decrease?

c) If we convert from traditional incandescent lighting to LED lights, which of the terms changes, and does this term increase or decrease?

d) If we switch from natural gas to coal, which of the terms changes, and does this term increase or decrease?

4. Consider this argument: "We cannot predict the weather in a week, so there is no way we can believe a climate forecast in 100 years." Is this argument right or wrong? Explain your answer.

5. If we emit significant amounts of carbon dioxide this century, how long will the planet remain warm?

6. Assume population grows at 2 percent/year and affluence grows at 3 percent/year.

a) How fast does the technology term have to decrease so that total emissions do not change?

b) How fast does the technology term have to decrease to reduce emissions by 20 percent in twenty years?

7. Explain how your level of wealth impacts how much emission of carbon dioxide you are responsible for.

8. In 2002, the Bush Administration set a goal of reducing greenhouse gas intensity by 18 percent by Year 2012. How ambitious is this goal? What does this goal tell us about changes in emissions?

# 9

# Impacts of climate change

Before the summer of 2010:

> Russia is a northern country and if temperatures get warmer, it's not that bad. We could spend less on warm coats.
>
> – Vladimir Putin, President of Russia[1]

After the summer of 2010:

> Practically everything is burning. The weather is anomalously hot. . . . What's happening with the planet's climate right now needs to be a wake-up call to all of us, meaning all heads of state, all heads of social organizations, in order to take a more energetic approach to countering the global changes to the climate.
>
> – Dmitri Medvedev, President of Russia[2]

You might have wondered as you read the first eight chapters, "OK, the climate is changing because of human activities. Why should I care?" In fact, warmer temperatures might sound good – you might associate them with fun things like vacations at the beach or summer cookouts. But reality is quite different. We rely in very important ways on the stability of the climate for things such as food and fresh water. Most people do not notice this reliance because it has been obscured by two centuries of scientific, technological, and economic advancements.

Nevertheless, it is there. And every once in a while, an event comes along that reminds us of the impact of climate on our lives. The heat wave of the summer of 2010 was one of those events for the Russians. The Russians learned the hard way that warmer temperatures do not mean tank tops and grilled hot dogs but instead mean wildfires, loss of agricultural crops, and suffering.

In this chapter, I cover the impacts of a changing climate. By the end, I hope that you recognize that climate change is a significant risk that we ignore at our peril.

## 9.1 Why should you care about climate change?

In Chapter 8, we saw that if the world does nothing to address climate change, we can expect global average temperatures to increase by a few degrees Celsius during the twenty-first century. This may not seem like much warming to you. After all, in

---

[1] Quoted in "Nyet to Kyoto, Blow for Campaign as Putin Jokes about Global Warming," *The Mirror,* September 30, 2003, p. 4.

[2] Quoted in "Will Russia's Heat Wave End Its Global-Warming Doubts?" *Time Magazine,* August 2, 2010.

many places summer days are 50 °C warmer than the winter days and daytime can be 25 °C warmer than the following night. And one day can be several tens of degrees Celsius warmer or cooler than the next. If you consider the size of these temperature variations, a change in the global average of a few degrees may sound insignificant.

In this case, however, your intuition is wrong. Although the temperature in any single place can vary considerably by season, by day, and even within a day, the variations tend to cancel when averaged over the entire globe. When you are experiencing the warmth of daytime, someone on the other side of the globe is experiencing the coolness of night. When it is summer where you live, it is winter in the other hemisphere. Heat waves in one location are generally canceled by a cold spell somewhere else (e.g., Figure 2.1). In other words, the large temperature variations you experience are nearly completely canceled by opposite variations somewhere else on the Earth.

Because of this cancellation, the global average temperature of the Earth is very stable, with Figure 2.2a showing year-to-year temperature variations of a few tenths of a degree. Moreover, seemingly small changes in global average temperature are associated with significant shifts in the Earth's climate. For example, the global annual average temperature during the last ice age was about 5 °C colder than that of our present climate. At that time, the Earth was basically a different planet: Glaciers covered much of North America and Europe, leading to a very different distribution of ecosystems, and because so much water was tied up in glaciers, sea level was approximately 120 m lower than it is today.

And during the summer of 2003, a heat wave struck Europe in which the average temperature in Europe was 3.5 °C above average. Despite this seemingly small amount of warming, this heat wave caused the deaths of several tens of thousands of people. And temperatures today are perhaps 1 °C warmer than they were a few hundred years ago, a period whose climate was different enough that it has been dubbed the Little Ice Age. Thus, we should take projections of a few degrees of warming seriously.

Furthermore, it is not just the size of the warming but the rate of warming that is of concern. It took more than 10,000 years for the planet to warm 5 °C and emerge from the last ice age – an average rate of 0.05 °C per century. The rate of warming predicted for the twenty-first century is a few degrees per century – about 100 times faster. Rate matters because the faster the warming occurs, the less time people and natural ecosystems have to adapt to the changes. If the sea level rises 1 m in 1,000 years, it seems likely that we could adapt gracefully to that change. But a 1-m increase in sea level in a century would be much harder to adjust to. And a 1-m increase in a decade would be a disaster, displacing millions of people and destroying trillions of dollars of infrastructure.

Another argument often made is that a warming of a few degrees should not cause concern because the Earth has gone through such warmings and coolings many times during its 4-billion-year history. This is undoubtedly true, as was discussed in Section 2.2. However, modern human society, with a population of several billion people, metropolitan areas with tens of millions of people, and reliance on industrial farming and large-scale built infrastructure is only a century or so old. Over this time, the Earth's climate has been stable, varying by less than 1 °C. Human society as we know it has never had to face several degrees of warming in a century. Thus, the argument that we have experienced this type of warming before is fundamentally misleading.

Finally, you might be asking, "How do I know that a warmer climate will not be better?" The reason it will not be is because both human society and natural ecosystems have adapted to our present climate. If the climate changes, then we will overall be less well adapted to our environment. As an analogy, imagine that you go to the tailor and get a suit fitted exactly to the shape of your body. At that point, no change in your body shape will improve the way the suit fits – for example, either gaining or losing weight will cause the suit to fit less well. In a similar fashion, any changes in the climate, either warming or cooling, will result in overall negative outcomes for human society.

As an example of how we rely on the climate, many structures in Alaska are built on permafrost (ground that remains frozen year round) with the implicit assumption that the ground will remain frozen. As long as that is true, the structures are stable. If the permafrost melts, however, the ground softens and can shift, potentially destroying structures (e.g., houses, bridges, roads) built on it. Thus, building on permafrost is a classic example of relying on an unchanging climate. Unfortunately, given the warming in the Arctic over the last century, assuming permafrost remains frozen is turning out to be a bad bet.

Agriculture provides another example of adaptation to our present climate. We farm where the climate provides suitable growing-season temperature and precipitation. Around these farms we build essential infrastructure to support agriculture: grain silos, processing plants, tractor dealers, seed suppliers, and cities for all of these people to live. If the climate shifts and the temperature and precipitation are no longer conducive to farming, all of these investments to support agriculture will no longer be useful. We may have to abandon them and rebuild the infrastructure in whatever region becomes conducive to farming.

As a final example, imagine that you build a marina on the edge of a lake. As soon as you pour the foundation, you are optimized for that particular lake level. If the lake level goes up, your marina floods. If the level of the lake drops, then you also face problems. If the drop is not too much, you might be able to adapt with small changes like lengthening the pier. But if the lake drops enough, then you must give up the original site and rebuild the marina closer to the new edge of the water. In the Western United States, where water levels are dropping in response to warmer temperatures, increased demand, and drought, this process of following the receding edge of the lake even has a name: "chasing water."

Not every single change in every region will be negative. Reductions in extreme cold events will have some benefits: less cold-weather mortality, benefits to agriculture of fewer freezing events (which can destroy some crops). Plant growth may well be enhanced in some regions due to higher carbon dioxide abundances and increased water availability. But these positive effects are expected to be outweighed by the more pervasive negative effects.

The upshot of this discussion is that, when it comes to climate, *change is bad*.

In this chapter, I will break the climate impacts into two components. I will first discuss the physical impacts on the climate system: how temperature, precipitation, sea level, extreme events, and other such phenomena will change. I will then discuss

## Annual mean surface air temperature change

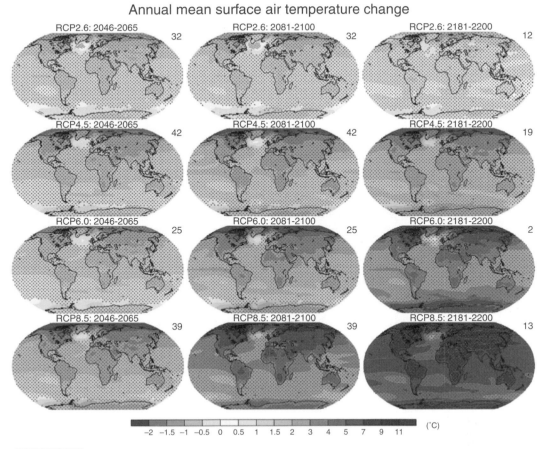

**Figure 9.1** The distribution of annual-average warming in the middle of the twenty-first century (left-hand column), the end of the twenty-first century (center column), and the end of the twenty-second century for the four RCP scenarios. Temperature increases are relative to the 1986–2005 average. These are calculated from an ensemble of climate models, with the number of models indicated in the upper right corner of each panel. Hatching indicates regions where there is disagreement among the models on the sign of the change, while stippling indicates regions where the models all agree on the sign of the change. This figure is adapted from figure 12.11 in Collins et al. [2013]. (See Color Plate 9.1.)

the impact of these changes on humans and those aspects of the environment that we rely on and care about.

## 9.2 Physical impacts

### 9.2.1 Temperature

Although the global average temperature is currently increasing and will almost certainly continue warming from each decade to the next, the increase is not uniform across the globe. Figure 9.1 shows the distribution of warming predicted for the various RCP scenarios, and there are several key features in the warming distribution. First, continents warm more than oceans because of the larger heat capacity of the oceans. As a result, warming in northern North America and Eurasia is projected to be more than 40 percent greater than the global average warming. Second, high latitudes will warm more than the tropics. This is primarily due to the ice-albedo feedback: The warming causes loss of ice, and the loss of ice exposes dark ocean, which absorbs more sunlight and leads to further warming (Figure 6.8). The models also predict more warming in the Arctic than in the Antarctic. These changes are all continuations of trends that have been observed over the past century (e.g., Figure 2.2b).

In general, adding greenhouse gases to the atmosphere tends to reduce temperature contrasts. We see this in the enhanced warming in the Arctic, which tends to reduce the temperature difference between the tropics and polar regions. We also expect more warming in winter than in summer (reducing summer-winter temperature contrasts), and more warming at night than during the day (reducing day-night temperature contrasts). To understand why this happens, you need to recognize that temperature variations in our climate are generally caused by variations in the distribution of sunlight. The polar regions, nighttime, and winter are all colder because they receive less sunlight than the tropics, daytime, and summer. The atmosphere also heats the surface, but because greenhouse gases tend to be well mixed in our atmosphere (because of their long residence times), the heating from greenhouse gases occurs evenly over the entire surface of the planet – it is the same at night as during the day, the same in winter as in summer, the same at high and low latitudes.

As the abundance of greenhouse gases increases, heating of the surface from greenhouse gases becomes stronger while heating from sunlight remains about the same. Thus, variations in solar heating with latitude, time of day, and season become a smaller component of the total heating of the surface, and will therefore lead to smaller temperature variations. In the limit of a planet such as Venus, with a massive greenhouse-gas-rich atmosphere, the heating of the surface by the atmosphere is roughly 16,000 $W/m^2$, which is about eighty times larger than solar heating. As a result, the variations in the solar input are so small in comparison that the temperature everywhere on Venus is approximately the same: 735 K.

Figure 9.1 shows that, in the middle of the twenty-first century, all of the RCP scenarios show about the same amount of warming. This small difference reflects the large inertia in our climate system from the large heat capacity of the oceans, as well as inertia in our economy – any policy to address climate change at reasonable cost will phase out fossil fuels over a few decades (phasing them out immediately would be very, very expensive). As a result, the warming over the next few decades is already determined by past emissions and our present mix of energy technology.

By the end of the twenty-first century, however, the scenarios have diverged: global average warming in the RCP8.5 scenario is four times larger than in the RCP2.6 scenario and twice that found in the RCP4.5 scenario. And the differences are even larger at the end of the twenty-second century. Thus, while our actions will not have much effect for the next few decades, policies to address climate change allow us to avoid the very large warming predicted for the second half of the twenty-first century and beyond.

Overall, we can have high confidence in the general shape of these predictions. There is, of course, uncertainty in the exact magnitude of the warming, both from

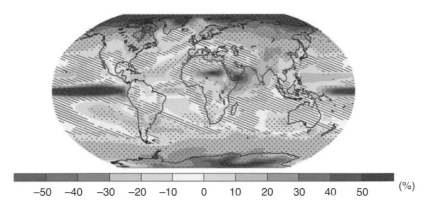

**Figure 9.2**   Change in annual mean precipitation over the twenty-first century as predicted by climate models driven by the RCP8.5 high-emissions scenario. The change is the percent change of 2081–2100 precipitation relative to the 1986–2005 precipitation. This is the average of thirty-nine models' predictions. Hatching indicates regions where there is disagreement among the models on the sign of the change, while stippling indicates regions where the models all agree on the sign of the change. (adapted from IPCC 2013, figure SPM.8). (See Color Plate 9.2.)

our understanding of the climate system and from uncertainty in how emissions will change in the future. But we can be confident that temperatures will continue to increase in the future and that the general distribution of the warming will be in accord with these model predictions.

## 9.2.2 Precipitation

As greenhouse gases increase, $E_{in}$ for the surface increases because of increased infrared radiation from the atmosphere falling on the surface. This leads to an increase in evaporation from the oceans, and because precipitation must balance evaporation, precipitation also increases. More quantitatively, total global precipitation is projected to increase by a few percent for every degree Celsius of global average warming.

Although total rainfall is expected to increase, the increase will not be distributed evenly. Predictions of changes in annual average precipitation from a set of climate models are shown in Figure 9.2. There is a large-scale shift of precipitation to higher latitudes, causing a decrease in many parts of the tropics. A general rule of thumb is that wet places will get wetter, while dry places will get drier.

In addition to changes in the amount of precipitation, there will also be shifts in the form. Less wintertime precipitation will fall as snow and more will fall as rain. This is more important than it might sound: When snow falls, the water does not run off until the snow melts in spring. Rain, on the other hand, runs off immediately, so changing the form of precipitation will change the timing of runoff, which has important implications for water availability.

These predictions of future precipitation patterns (like Figure 9.2) are less certain than the temperature predictions (like Figure 9.1). Precipitating clouds can be small (sometimes just a few kilometers across), and climate model grids are too coarse to directly simulate such small structures. In the end, we can have some confidence in the predictions at a qualitative level (e.g., shifts in precipitation location, more rain falling in heavy events) but confidence should be lower for the quantitative details of the spatial distribution of the changes.

### 9.2.3  Sea-level rise and ocean acidification

Sea-level rise is one of the most certain impacts of climate change. As we learned in Chapter 2, the sea rises in response to warming temperatures for two reasons. First, as grounded ice melts, the melt water runs into the ocean, increasing the total amount of water in the ocean and, therefore, sea level. Second, like most things, water expands when it warms, which also tends to raise sea level. Measurements (Figure 2.10) confirm that sea levels have indeed been rising as temperatures have gone up, and we can be certain that the seas will continue to rise into the next century.

The latest IPCC report predicts that sea level will rise 45 to 75 cm (18 to 30 inches) above today's levels by 2100. Even the low end is nearly three times the sea-level rise of the twentieth century of 17 cm (7 inches). But even more worrying is that we have strong evidence from the last interglacial (about 120,000 years ago), when temperatures were about 2°C warmer than today, that sea levels were at least 5 m (16 ft) higher, mainly due to melting of the Greenland ice sheet. Given that we are likely to see this amount of warming above preindustrial temperatures during the twenty-first century, why are the estimates of twenty-first-century sea-level rise (relatively) small?

The reason is that it takes a very long time – many centuries, or even millennia – to melt this much ice. This means that warming temperatures this century and beyond will likely guarantee a few meters of sea-level rise, but we will only see a small fraction of this this century. Most of the increase in sea level due to our emissions will occur far in the future. Put another way, our actions today might commit residents of the Earth in the year 2500 or 3000 to live in a world with much higher seas.

Ocean acidification is another certain consequence of continued emissions of carbon dioxide. As we explored in Chapter 5, a significant fraction of carbon dioxide emitted to the atmosphere by humans ends up in the oceans. In the liquid environment of the ocean, carbon dioxide reacts with water and is converted into carbonic acid (Equation 5.3). The net result is that, as the oceans absorb more and more carbon dioxide, the oceans will become more and more acidic.[3] In fact, since the industrial revolution, the absorption of carbon has lowered the ocean's pH by approximately 0.1. At present, the ocean is now more acidic than it has been for 20 million years.

### 9.2.4  Loss of ice

Ice melts reliably at 0°C, so we expect the planet's ice to vanish as the globe warms. Indeed, we have observed a steady retreat of the world's glaciers (Figure 2.6), Arctic sea ice (Figure 2.7), and the Greenland and Antarctic ice sheets over the past few decades (Figure 2.8). This is one of the main contributors to the observed increase in sea level.

---

[3]  The present pH of the ocean is approximately 8, meaning that it is a base. Acidification here means that the pH is decreasing, not that the ocean will actually become acidic. For the ocean to actually become acidic, its pH would have to drop below 7, which is very unlikely.

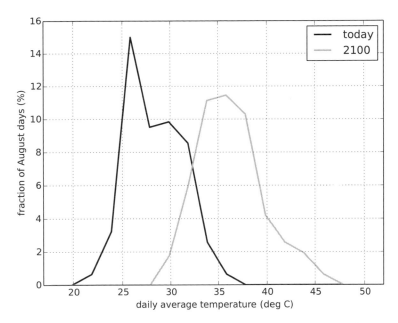

Figure 9.3 The frequency of occurrence of daily average temperatures around Dallas, Texas, in two periods: 2006–2015 and 2091–2100. Temperatures come from the GFDL-CM3 climate model driven by the high-emission RCP8.5 scenario.

In the future, we can expect sea ice to continue to decrease in the Arctic, and to begin to decrease in the Antarctic as temperatures there begin to increase. Most predictions are that the Arctic ocean will be entirely ice free during summertime at some point in the twenty-first century, although exactly when is still being debated.

### 9.2.5 Extreme events

Extreme weather events – heat waves, droughts, floods, hurricanes, tornadoes, to name just a few – are some of the most consequential ways that the environment affects us. If a warming climate causes increases in the frequency or severity of these types of events, this would be a scary and consequential outcome of the burning of fossil fuels.

Some changes in extremes are certain, or nearly so. We can be certain that extreme heat events will become more frequent and severe. The impacts of this are serious, as evidenced by the tens of thousands of people who perished during the European heat wave in the summer of 2003. In fact, extreme heat is the largest weather-related cause of death in the United States.

Figure 9.3 shows a climate model calculation of how temperature extremes might change in a warmer climate under a high-emissions scenario (RCP8.5). The panel shows the distribution of daily average August temperatures around Dallas, Texas, for the periods 2006–2015 and 2091–2100. The daily average temperature increases between these two periods from 28°C to 36°C – an increase of 7.8°C (14°F). Figure 1.1 showed a similar, although much smaller, shift in the frequency distribution of daily temperatures in Fairbanks, Alaska, in response to the warming of the twentieth century.

We can estimate the occurrence of today's extreme temperatures as the temperature exceeded by the warmest 5 percent of days between 2006 and 2015. In the climate model, this corresponds to days with average temperatures above 32.4°C. In the 2091–2100 period, temperatures above 32.4°C are projected to occur on 78 percent of the days – a dramatic increase in extreme temperatures.

The hottest day of the 2006–2015 period had a daily average temperature of 36.5°C. Disturbingly, 39 percent of the days between 2091 and 2100 will be hotter than this. This means that the climate is reaching uncharted territory: temperatures that never occur today will be occurring frequently by the end of this century. We can verify this trend in the historical record, where we see record heat more often than record cold. In 2012, for example, there were 34,008 daily high temperature records in the United States and 6,664 daily low temperature records.[4]

In addition to changes in the pattern of precipitation (Figure 9.2), it is likely that more rainfall will come in the heaviest downpours, which continues a trend observed over the last few decades. During a heavy downpour, the soil saturates before the end of the rain event, and the remaining rain therefore runs off, leading to a number of negative consequences, such as increased risk of flooding and loss of freshwater for use by humans and ecosystems.

An increase in the fraction of heavy events also tends to increase the time between rain events. Combined with warmer temperatures, which will increase the rate at which water is lost from soils by evaporation, this will also contribute to increased drought occurrence. Thus, we get the surprising result that both wet and dry extremes will grow more likely in the future: wet extremes, with associated risks of flooding, increased erosion, and landslides; and dry extremes, with associated risks of water shortages and drought.

One topic that is frequently talked about is hurricanes, typhoons, and cyclones (these storms are the same; the name is determined by which ocean the storm appears in). These are some of the most dramatic weather events we face, and they can cause enormous damage. So how will hurricanes and their impacts change in the future? First, we can say with certainty that the impacts of hurricanes will be more severe because a rising sea level will lead to higher storm surges and flooding, as well as increased rates of coastal erosion. As far as the storms themselves go, research indicates that hurricanes may get stronger, although the number of these storms may decrease. But our confidence in this impact is far from certain.

## 9.3 Impacts of these changes

As described in the previous section, climate change will bring about a set of certain impacts (e.g., increasing temperatures including extreme heat events, changes in

---

[4]  Numbers from www.climatecentral.org/news/noaa-2012-was-warmest-and-second-most-extreme-year-on-record-15436. See also Meehl et al. (2009), Relative increase of record high maximum temperatures compared to record low minimum temperatures in the U.S, *Geophysical Research Letters*, 36, L23701, doi:10.1029/2009GL040736.

precipitation, increases in sea level, increasing acidity of the ocean) as well as a large number of potential impacts (e.g., increasing flood and drought intensity and frequency, increases in hurricane intensity). These physical changes in the climate system are only the first step in determining climate impacts. Determining impacts also requires an understanding of how vulnerable human and natural ecosystems are to these changes.

One thing that makes this a challenging exercise is that humans can adapt to a changing climate. For example, many buildings in Boston are currently not air-conditioned, because there are only a few days a year that are hot enough to require cooling. As the climate warms, air-conditioners can be installed in buildings as needed to accommodate the heat. Similarly, increases in sea level can be dealt with by building seawalls or by relocating people. And many ecosystems that humans rely on are intensively managed, such as agriculture, commercial forests, rangelands, and fisheries. Because human management dominates these systems, the ability to adapt management practices to changing conditions offers the possibility of mitigating at least some of the harmful impacts. At the same time, disruption of these systems by climate change may have severe human impacts because we depend on them so much.

Agriculture is probably the most obvious economic sector that is going to be impacted by climate change – through changes in temperature (photosynthesis in some of our most important crops is most efficient at temperatures between 20 °C and 25 °C) and fresh water availability (a good rule of thumb is that it takes 1,000 tons of water to produce 1 ton of grain, and about 15,000 tons to produce 1 ton of meat).[5] Adjusting farming practices can alleviate most of the negative impacts for local warming below about 3 °C. Above this threshold, food production is projected to decrease. As Figure 9.1 shows, the continents warm faster than the ocean, so 3 °C local warming in agricultural areas might occur as early as the middle of the twenty-first century in the high-emission scenarios. A decrease in food production beginning in just a few decades would constitute a massive challenge given the projected increase in population during the twenty-first century.

This is just one example of how changes in the climate will affect us. In particular, changes in fresh water will be a major challenge of climate change because we rely on it for so many things. For example, most electricity power plants require huge amounts of water for cooling, which is why these plants are usually situated on rivers or lakes. During droughts, low water levels sometimes force these plants to shut down.

In some regions, overall water availability is expected to increase. For example, river runoff is projected to increase by 10 percent to 40 percent by mid-century at higher latitudes and in some wet tropical areas, including populous areas in East and Southeast Asia. However, the beneficial impacts of increased annual runoff in these areas may be tempered by changes in the timing of the runoff. For example,

---

[5] One frequently hears that increasing carbon dioxide will be good for plants – i.e., "carbon dioxide is plant food." Atmospheric carbon dioxide is indeed a key ingredient in plant growth and, everything else being equal, more carbon dioxide in the atmosphere would be expected to increase the rate of plant growth. However, other changes, such as changes in temperature and precipitation, are expected to offset the benefits of increased carbon dioxide, particularly for warming of more than a few degrees Celsius.

summertime runoff from melting snowpack provides an important source of fresh water to the U.S. Pacific Northwest at a time when there is little rainfall. Warming temperatures, however, will lead to less wintertime precipitation falling as snow and more as rain, and the snow that does fall will melt earlier. Both of these effects will tend to shift runoff from the summertime, when the water is most needed, toward winter and spring, when it is less needed. This will increase stresses on summertime water supplies.

In the mid-latitudes and dry tropics (e.g. the Mediterranean Basin, Western United States, southern Africa, and northeastern Brazil), decreases in rainfall and increases in temperature will lead to a significant decrease in water resources. Drought-affected areas in these regions are expected to increase in extent. Humans and human-managed ecosystems should be capable of adapting to decreases in fresh water supplies, as long as the decreases are modest. Large reductions in fresh water supplies associated with the highest warming scenarios, however, would likely exceed our ability to respond gracefully.

Other impacts of climate change will be much harder to manage. Many of these are on systems that humans do not manage because they are fundamentally unmanageable, such as rising sea level. As discussed earlier, sea level is predicted to rise 45 to 75 cm (18 to 30 inches) above today's levels by 2100. This is a huge challenge for human society because a significant fraction of the world's population lives within a few feet of sea level. Moreover, some of the world's most productive farmland is located in river deltas and other regions that are particularly sensitive to sea-level rise.

Even small amounts of sea-level rise will therefore have significant negative implications. In Florida, for example, a sea-level rise in the middle of the projected range would inundate 9 percent of Florida's current land area at high tide.[6] This includes virtually all of the Florida Keys as well as 70 percent of Miami-Dade County. Almost one-tenth of Florida's current population, or nearly 2 million people, live in this vulnerable zone, and it includes residential real estate now valued at over $130 billion. It also includes important infrastructure, such as two nuclear reactors, three prisons, and sixty-eight hospitals. And this is just Florida. Multiply these impacts to account for all of the places on the planet where people live near sea level, and you can get a feel for how big a problem this is going to be.

And regions do not actually have to be submerged to be affected. Increased sea level will increase the frequency of flooding from extreme sea-level events, so that a flood event that occurred, say, every 100 years, may occur every few years by the end of the century. As the flooding of New Orleans after Hurricane Katrina showed, these events cause significant loss of life as well as economic destruction.

It is worth noting that sea-level rise provides a particularly good example of how we are adapted to our present climate. When cities such as New Orleans or Miami were founded, the original inhabitants did not take into account the possibility of sea-level rise. Rather, they assumed that the sea level would remain pretty much as they found it, so they built their city in concert with that particular sea level. Any change in sea level, either up or down, has negative effects on these cities.

[6] See Stanton and Ackerman (2007).

Humans can adapt to rising sea levels, of course, but all of the adaptation solutions are hard and expensive. Our basic choice is between building extremely expensive infrastructure to protect the city (e.g., sea walls) or simply abandoning those areas and the trillions of dollars of infrastructure in them. Given the certainty of sea-level rise, the most rational approach would be to immediately cease building new infrastructure in regions that are likely to be overrun by sea-level rise in this century (i.e., Miami) and instead begin an orderly multi-decadal retreat from the coast. That, of course, is not happening.

Another unmanageable impact of climate change is the acidification of the ocean. Many ocean organisms build shells or skeletons out of calcium carbonate, and their ability to do this is affected by the acidity of the ocean. As the ocean becomes more acidic, these species will at first find it more difficult to extract carbonate from the water for use in their shells or skeletons. Eventually, the acidity will increase to the point where it is fatal for the species. It is important to realize that ocean acidification is not just a theory – it has happened before. During the PETM (discussed in Section 7.5), a massive amount of carbon was emitted into the atmosphere, which subsequently dissolved into the ocean. That event was accompanied by an acidification of the ocean that dissolved much of the carbonate sediment there (as shown in Figure 8.6).

Some species will adapt better to increasing acidity than others. As a result, the mix of species in ocean ecosystems will shift, resulting in new and novel arrangements of species. This will affect humans because we rely in important ways on the ocean: e.g., about a billion people rely on the ocean as their primary source of protein. And, like sea-level rise, this is basically an unmanageable impact. If the amount of protein available for human consumption from the ocean decreases, there are no simple adaptations to solve that problem – the protein will have to be made up elsewhere or people will starve.

Natural ecosystems and their constituents will also be affected. At the individual species level, for example, research shows that warming temperatures are presently driving lizards to extinction. During spring, when energy demands are highest because lizards are reproducing, the warming temperatures reduce the amount of time that lizards can forage for food (if the temperatures get too high, cold-blooded lizards have to rest). If temperatures continue to increase (which we expect), then at some point the time available to look for food diminishes to the point where lizards simply cannot find enough food – and extinction ensues. This is already happening, and extrapolating into the future, global warming may lead to the extinction of 40 percent of all global lizard populations by 2080. Note that it is not the global average temperature that matters to lizards, or even the local average temperature, but the local daily temperatures during one particular time of the year. This emphasizes that it is the details of climate change that ultimately matter, not the broad-brush changes in global average quantities.

Now you may not care much about lizards, but you should care about their extinction for two reasons. First, the environment is a tightly coupled system. There are many examples in history in which humans have intentionally removed a species from the environment because they thought it was harming them (e.g., getting rid of birds because the birds were eating crops) only to find out that that change led to more problems than it solved (e.g., the birds were also eating insects, and with the birds gone the insects proliferated and destroyed much more of the crops than the birds were eating). Today's modern world obscures many of these relationships, but they nonetheless still exist. Removing lizards from an ecosystem may have important effects on the rest of the environment that we do care about, just like pulling a single thread on a sweater can unravel the entire thing.

Second, this is not just about lizards. As warming temperatures drive lizards to extinction, the same warming temperatures will be having deleterious effects on many other species. In fact, a significant fraction of plant and animal species may be at increased risk of extinction if global average temperatures increase by a few degrees Celsius. Thus, at the expense of a mangled metaphor, lizards may be the canary in the coalmine.

Changes to individual species will project onto changes in entire ecosystems, such as alpine meadows or temperate forests. As the climate changes, each component species of an ecosystem will be affected in its own way. Some species may adapt readily, whereas others may be unable to adapt fast enough to survive. Species will also be subject to human interventions and constraints such as land-use change, barriers, and intentional or inadvertent transport.

The aggregate result will be that ecosystems will evolve, with new relationships among incumbents and new arrivals developing in each location. In some cases, the new assemblages may be similar enough to present ecosystems that we can think of them as basically unchanged. In other cases, however, the new systems may be unlike any present ecosystems, with new species and relationships between them and other ecological surprises. Some ecosystem types are likely to be lost entirely, such as alpine systems, coastal mangrove systems, and coral reefs.

---

### An aside: What are *ecosystem services*?

Natural ecosystems provide enormous benefits to human society. The mangrove forests that grow in shallow salt-water coastal regions are good examples. They provide important protection for coastal areas from erosion, storm surge (especially during hurricanes), and tsunamis. Their loss will cost us money – either we will have to build expensive coastal defenses to replace the natural defense provided by the mangrove or we will have to absorb the cost of increased coastal damages. This value, provided to humans for free, is what we mean when we talk about ecosystem services.

Another good example is pollination by bees. Many crops (e.g., apples, almonds, blueberries) are directly dependent on bee pollination as part of their growing cycle, and the total value of these pollinated crops in 2010 was $16 billion. The cost of pollination (by wild bees, at least) is provided free of charge by nature to us. In China, a decline of wild bees has forced farmers to hire people to go from flower to flower and hand-pollinate the flowers using tiny brushes. Thus, ecosystems provide important economic benefits to our society; the impacts on them from climate change will therefore impose potentially steep costs on society.

---

A particularly important factor in determining the severity of the impacts on natural ecosystems will be the rate of climate change. Ecosystems have adapted to large climate change in the past, such as the warming from the previous ice age. However, the warming predicted for the next century will be incredibly fast – perhaps one hundred times faster than the average rate of warming since the last glacial maximum 20,000 years ago. As the rate of warming goes up, the ability of the environment to gracefully adapt to the changes declines. What is uncertain is how much less gracefully, and with what consequences.

We can also expect human health to be negatively impacted by climate change. Some of these health impacts follow closely from changes already discussed, such as negative health consequences of warmer temperatures or malnutrition associated with reductions in food availability. In addition to those impacts, we expect warmer, more humid days to enhance the photochemical reactions that cause air pollution, leading to more smoggy days as the climate warms, along with the associated health impacts.

Warming temperatures also increase disease risk as a result of expansions in ranges of animals that transmit the diseases (e.g., mosquitoes), shortening of the diseases' incubation periods, lack of very cold temperatures that can kill the transmitters, and disruption and relocation of large human populations. Moreover, increases in water temperature, precipitation frequency, and other factors could increase the incidence of water contamination with harmful pathogens, resulting in increased human exposure.

However, because human health is part of an intensively managed human health care system, we expect human society to adapt and effectively manage many of the impacts – for modest warming, at least. As the warming gets progressively larger, the ability and effectiveness of strategies to manage the health threats from climate change will decrease and the net negative impact on human society will grow rapidly.

It should also be noted that intensively managing climate change costs money. This means that the ability to adapt is not evenly spread across the globe. Rich, well-governed places such as the United States or Europe have resources that can be applied to adapting to climate change. For small climate change (i.e., the bottom end of the range in Figure 8.5), rich countries will likely find most effects of climate change to be manageable without too much social disruption. If climate change falls toward the upper end of the predicted range in Figure 8.5, then climate change is expected to be a serious, perhaps insurmountable challenge for even these rich countries.

Approximately 2 billion people, however, are so crushingly poor that they have no additional resources available to address even minor climate change. For them, installing air-conditioners or building coastal defenses in response to rising seas, developing new freshwater infrastructure in response to water shortages, improving public health infrastructure in response to a new disease outbreak, and the like are simply not affordable options. For these poorest of the poor, even small climate change will be a major challenge. If climate change falls toward the upper end of the predicted range in Figure 8.5, then it would be a certain disaster for the poorest.

Because some countries will be unable to manage even moderate climate change, many experts view climate change as a potentially destabilizing social force. Wars over resources (e.g., fresh water) or large-scale migration to escape climate extremes (e.g., droughts) could cause a range of social and economic disruptions, leading to failed states and the associated impacts on regional and world stability. This has led national defense analysts at the Pentagon to categorize climate change as a threat to U.S. national security.

# 9.4 Abrupt climate changes

Many of the changes I just described – changes in temperature, precipitation patterns, sea level, and so on – are steady changes in the climate system. For example, we expect the climate to warm by a few tenths of a degree Celsius per decade (as suggested by plots such as Figure 8.5), and we expect sea level to rise by a few centimeters per decade. These changes, while incredibly fast geologically, are gradual on human timescales.

An *abrupt climate change* is a sudden and significant shift in some aspect of the climate. As an analogy, imagine that you are sitting in a canoe and you start to lean over. At first, the canoe tilts with you – until, that is, you pass a critical threshold and the canoe suddenly flips over, throwing you and everything else in the canoe into the river. That is an abrupt change.

For the climate, the worry is that the climate will not warm smoothly as greenhouse gases are added to it. Rather, we will add enough greenhouse gas that the climate system will undergo a large and rapid shift to an entirely new climate state – equivalent to the canoe rapidly transitioning from right side up to upside down. In the case of climate, the large climate shift might occur on a timescale of decades.

This possibility concerns scientists because abrupt changes have happened in the past. During the PETM approximately 55 million years ago, there was a rapid release of greenhouse gases and a subsequent warming of 5–9 °C in just a few thousand years. It is not known what caused the release of greenhouse gases, but one possibility is that it was due to a carbon cycle feedback – for example, an initial warming melts permafrost, leading to the release of carbon stored in it, which leads to more warming, and so on. Scientists worry that warming due to humans could do the same thing: melt permafrost and release massive amounts of carbon dioxide and methane. That would doom the planet to another PETM.

In addition, roughly 12,000 years ago, as the Earth was emerging from the depths of the last glacial maximum, the temperature suddenly plunged (at least in the mid-and high latitudes of the northern hemisphere). The period of low temperatures during the millennium that followed, today known as the Younger Dryas, is thought to have been due to a massive release of water into the North Atlantic from melting glaciers. This freshwater influx disrupted the ocean currents, in particular the Gulf Stream. Because the Gulf Stream transports heat from the tropics to the high latitudes, the shutdown of the Gulf Stream caused mid- and high-latitude temperatures to plummet (this was the basic scientific premise behind the movie *The Day After Tomorrow*).

Thus, abrupt changes do happen and we must take their possibility seriously. However, beyond acknowledging the possibility, there is little the scientific community can say. Climate models do not predict the occurrence of an abrupt climate change, and most experts view the probability to be low, but not zero, over the coming century. If an abrupt change did occur, though, it could be a catastrophe. Such low-risk, high-consequence events pose significant challenges to our society. The tendency is to ignore the risk until it occurs, which is why dams are built after floods, and not before. However, for these types of events, once the abrupt change takes place, it will be very difficult, if not impossible, to gracefully manage. This makes the strategy of ignoring the risk a precarious proposition.

It is also possible that abrupt impacts can occur in a slowly changing climate when the climate reaches a threshold set by human vulnerability. A good example is the flooding of lower Manhattan during Superstorm Sandy. As the storm surge reached certain levels, abrupt and large impacts were suddenly felt. For example, as long as the water level stayed below the entrance to the subways, there was no flooding of the subway system. But as soon as the water overtopped the entrance to one station, water rushed into the system and, traveling through the subway tunnels, was able to flood large parts of the system. Some sections were out of commission for months, and the cost of fixing it was immense.

# 9.5 Chapter summary

- The amount of warming predicted for the twenty-first century (a few degrees Celsius) is comparable to the warming since the last ice age ($5\,°C$). This means that the warming over the next century or two may herald a literal remaking of the Earth's environment and our place within it.
- We are adapted to our present climate, so any significant change (in any direction) is likely to be detrimental.
- There are a number of virtually certain impacts of climate change: The climate will get warmer (with more extreme heat events), precipitation patterns will change, sea level will rise, and the oceans will become more acidic. These are serious impacts that should compel our attention.
- There are also a number of more speculative changes in extreme events (e.g., floods, droughts, hurricanes). Many of these changes are likely, but we cannot be sure of the severity.
- These changes may have important negative impacts on humans. This includes impacts on agriculture and freshwater availability, as well as public health consequences.
- Natural ecosystems may also be severely disrupted. These natural systems provide services of great value to humans, so their disruption could provide significant challenges to us.
- The impacts of these changes on human society will not be distributed evenly. The wealthy countries of the world will have an easier time adjusting than will the poor countries of the world, and there may be some people who benefit, particularly if the warming is small. As the amount of warming increases, negative impacts will increasingly dominate the benefits, and even the richest countries will be severely challenged.
- Abrupt changes are low-probability, high-consequence events. An example of an abrupt change was the reorganization of the ocean's circulation during the Younger Dryas period about 12,000 years ago. Although scientists do not expect them to occur this century, they cannot be ruled out.

# Additional reading

Much has been written about the impact of climate change on humans and the rest of the environment. Here are a few notable works.

The IPCC's Working Group II focuses on impacts of climate change and our ability to adapt to them. You can download and read the IPCC's Fifth Assessment Report at ipcc-wg2.gov/AR5/report/.

The IPCC has also released an assessment of the links between climate change and extreme weather and climate events, the impacts of such events, and the strategies to manage the associated risks. See C. B. Field, V. Barros, T. F. Stocker, D. Qin, D. J. Dokken, K. L. Ebi, M. D. Mastrandrea, K. J. Mach, G.-K. Plattner, S. K. Allen, M. Tignor, and P. M. Midgley (eds.), *Managing the Risks of Extreme Events and Disasters to Advance Climate Change Adaptation*. A Special Report of Working Groups I and II of the Intergovernmental Panel on Climate Change (Cambridge and New York: Cambridge University Press, 2012), 582 pp. (download at ipcc-wg2 .gov/SREX/).

William Nordhaus, The Climate Casino: Risk, *Uncertainty, and Economics for a Warming World* (New Haven, CT: Yale University Press, 2013). This is an excellent book about the economics of the climate problem. Part II of the book contains a first-rate description of how economists evaluate climate impacts.

Here are a few other books I recommend that describe the impacts of climate change. They are all aimed at the general public, so they are easy reads and require no specialized knowledge: E. Kolbert, *Field Notes from a Catastrophe: Man, Nature, and Climate Change* (New York: Bloomsbury USA), 2006; M. Lynas, *Six Degrees: Our Future on a Hotter Planet* (Washington, DC: National Geographic, 2008); J. Diamond, *Collapse: How Societies Choose to Fail or Succeed* (New York: Penguin, 2011).

See www.andrewdessler.com/chapter9 for additional resources for this chapter.

# Terms

Abrupt climate changes
Ecosystem services

# Problems

1. Your third cousin once removed asks you why we will not be better off in a warmer climate. What do you tell him?
2. Your friend says, "Climate scientists are such alarmists. First they say that floods will become more frequent, and then they say that droughts will become more frequent. Come on, which one is it? They cannot both occur!" What do you tell her?

3. As discussed in this chapter, temperatures are not expected to rise uniformly across the globe.
   a) Why is there more warming at high latitudes than the tropics?
   b) Why will land warm more than the ocean?
   c) Why do temperature contrasts (e.g., night vs. day) decrease in a warmer climate?
4. Precipitation
   a) How is precipitation expected to change in a future climate?
   b) Why do changes in the form of precipitation (rain vs. snow) matter?
5. Explain a few ways that climate change impacts public health.
6. Why will it be easier for the United States and Western Europe to deal with climate change than countries in Africa?
7. Explain how climate change affects our national security.
8. a) What do scientists mean when they talk about "abrupt climate change"?
   b) Give an example of an abrupt climate change that has occurred in the past.

# Exponential growth

Before we continue our discussion of climate policy, we need to take a detour to examine *exponential growth*, which may be the most important term that you have never heard of. It touches many aspects of your life, from the growth of credit card debt and housing prices to governing key processes in biology, physics, economics, and, yes, climate change.

## 10.1 What is exponential growth?

First, a definition: *exponential growth* means that the rate of growth is directly proportional to the present size. A good example of exponential growth is the accumulation of money in a savings account. Imagine that you deposit $100 into a bank account with an *interest rate* of 10 percent/year.[1] After the first year, you receive interest equal to 10 percent of the balance of $100, which is $10. This interest raises the balance of the account to $110. After a second year, the interest is 10 percent of the balance of $110, which is $11. This increases the balance to $121. Table 10.1 shows the growth of the bank account over 101 years.

This growth is exponential because the increase in the bank balance during any year is proportional to the bank balance in that year. In fact, anything growing at "*x* percent/year" is growing exponentially.

The key parameter in exponential growth is the rate of growth, or $r$. For bank balances, credit cards, or mortgages, $r$ is usually called the interest rate, whereas in other contexts $r$ may have other names (later in the chapter, I will refer to $r$ as the discount rate). Usually, $r$ is expressed in percent/year. Given a growth rate of $r$ percent/year, an initial quantity $P$ will grow by a factor of $1 + r/100$ in one year. So, after one year, the quantity has grown to $P(1 + r/100)$. For a bank balance of $100 and an interest rate of 10 percent/year, the bank balance after one year is $100(1 + 10/100) = \$100(1.1) = \$110$, the same answer we obtained earlier.

At the end of two years, the balance is $P(1 + r/100)(1 + r/100)$. This is simply the balance at the end of the first year, $P(1 + r/100)$, multiplied by another factor of $1 + r/100$ to account for growth during the second year. Thus, the bank balance at the end of the second year is $\$100(1 + 10/100)(1 + 10/100) = \$100(1.1)(1.1) = \$110(1.1)^2 = \$121$.

---

[1] We are assuming here that the interest is compounded annually. Most bank accounts and credit cards calculate interest monthly, meaning that the balance is increased each month by the balance times the annual interest rate divided by 12. Monthly compounding grows the balance faster for a given interest rate than annual compounding.

| Table 10.1 Calculation of the balance of a bank account with an initial investment of $100 at an interest rate of 10 percent/year | | |
|---|---|---|
| Year | Interest ($) | Balance ($) |
| | | $100 |
| 1 | 10 | 110 |
| 2 | 11 | 121 |
| 3 | 12.10 | 133 |
| 4 | 13.30 | 146 |
| 5 | 14.60 | 161 |
| 6 | 16.10 | 177 |
| 7 | 17.70 | 195 |
| 8 | 19.50 | 214 |
| ⋮ | | |
| 100 | 125,278 | 1.38 million |
| 101 | 137,806 | 1.52 million |

You may well be able to see a pattern here. After $n$ years, an initial investment of $P$ will grow to a final value $F$:

$$F = P(1 + r/100)^n \qquad (10.1)$$

This is the formula I used to generate the values in Table 10.1.

## 10.2 The rule of 72

When thinking about exponential growth, it is frequently useful to consider the *doubling time* – the length of time that it takes for something growing exponentially to double. From Table 10.1, for example, we see that $100 invested at 10 percent/year will double in approximately seven years to $200. After another seven years, the $200 doubles to $400, and after another seven years, the $400 will grow to $800.

A simple way to estimate the doubling time is to use the *rule of 72:* The doubling time is 72 divided by the growth rate (in percent/year). Using this equation, we see that the doubling time at 10 percent/year is $72/10 = 7.2$ years, a result that is consistent with Table 10.1. Note that doubling time is a function only of growth rate. Thus, for a growth rate of $r$, the doubling time is the same regardless of the size of the growing quantity.

Using this rule, you can frequently do exponential growth problems with pencil and paper – or even in your head. For example, let us put $100 in the bank at 7.2 percent interest in Year 2000. What is the balance in Year 2100? The doubling time is $72/7.2 = 10$ years, so that the balance doubles every 10 years. Table 10.2 shows the balance at the end of every decade.

After 100 years the $100 investment has grown to $102,400 – illustrating the power of exponential growth. In equation form, an initial investment of $P$ has grown after $n$ doublings to a final value $F$:

$$F = P(2^n) \qquad (10.2)$$

| Table 10.2 The balance of $100 invested in Year 2000 at an interest rate of 7.2 percent/year | | |
|---|---|---|
| Year | No. of doublings | Value ($) |
| 2000 | – | 100 |
| 2010 | 1 | 200 |
| 2020 | 2 | 400 |
| 2030 | 3 | 800 |
| 2040 | 4 | 1,600 |
| 2050 | 5 | 3,200 |
| 2060 | 6 | 6,400 |
| 2070 | 7 | 12,800 |
| 2080 | 8 | 25,600 |
| 2090 | 9 | 51,200 |
| 2100 | 10 | 102,400 |

This is the equation I used to calculate the values in Table 10.2. And in a pinch, I could have done the calculation with just pencil and paper.

We could also have used Equation 10.1 to calculate the balance in 2100: $F = \$100(1 + 7.2/100)^{100} = \$104{,}587$. This is very close to the value calculated for 2100 by use of the doubling time, although the estimates differ slightly. The difference results from the fact that the rule of 72 is approximate, so calculations using it are almost always slightly off. In most cases, though, the rule of 72 is accurate enough.

As another example, imagine investing $100 at an interest rate of 14.4 percent. How long would you have to leave this investment in the bank to yield $1 trillion ($10^{12}$)? First, let us figure out how many doubling times it would take. Using Equation 10.2, we can write the relevant equation as

$$\$100(2^n) = \$10^{12} \tag{10.3}$$

Solving this equation[2], we find that $n = 33.2$. In other words, an initial investment of $100, doubled 33.2 times, yields $1 trillion. Now let us calculate how many years that takes. At an interest rate of 14.4 percent/year, the doubling time is $72/14.4 = 5$ years, so 33.2 doublings takes 166 years. The growth of this investment is plotted in Figure 10.1.

Most of the accumulation occurs in the last few doubling periods. In fact, $500 billion dollars, half of the total, is earned in just the last doubling period, and 97 percent of the $1 trillion is earned in the last five doubling periods (25 years). And, had the investment period been cut in half (from 166 to 83 years), you would have $10 million at the end, 1/10,000th of what you earn for the entire time period. Thus, the exponential growth is heavily weighted toward the very end of the investment period. This has some important consequences, which we investigate in the next section.

---

[2] There are several ways to solve this equation. One way is to iterate. This means that you guess a value for n and plug it into your calculator. If that n produces too large a number, then reduce your estimate of n and repeat the process; if it produces too small a number, then do the opposite. With a bit of practice, you can quickly converge on the correct answer with five to ten guesses. You can also solve the equation algebraically by taking the log of both sides of the equation and rearranging to solve for n.

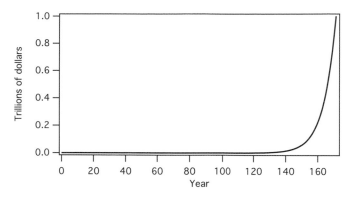

Figure 10.1   Value of $100 invested at 14.4 percent interest as a function of years invested.

## 10.3 Limits to exponential growth

Where I live, College Station, Texas, the population growth rate peaked at 11 percent/ year in the 1970s. At that rate, College Station's population was doubling every $72/11 = 6.5$ years, or about 15 doublings per century. At that growth rate, sustained over a century, College Station's population would have increased from about 37,000 in 1980 to 1.3 billion by 2080. This is an impossibly high population and means that this growth rate could not possibly be sustained for 100 years. In fact, considering the practical limits of city growth, such high population growth is unsustainable for much more than a decade or two. Thus, it should have been obvious in the 1970s that the population growth rate of College Station would decrease – and it has, to roughly 3 percent/year today.

A growth rate of 3 percent/year sounds much more sustainable, and it is, but exponential growth is so stunningly fast that even seemingly low growth rates can be unsustainable for more than a few centuries. For example, at 3 percent/year, College Station's population would approach 1 billion in about 300 years or so. This is also clearly unattainable in any practical sense, and it says that, over the upcoming centuries, the population growth rate of College Station must decline even further.

This brings us to the first important rule of exponential growth, best expressed by economist Kenneth Boulding, "Anyone who believes exponential growth can go on forever in a finite world is either a madman or an economist."[3]

A second rule of exponential growth is that, when the end comes, it comes quickly. As an example, let us consider a resource consumed by humans, such as water. Let us assume that in Year 2000, a community is consuming 100 million gallons of water per year. As the community grows, the rate of consumption doubles every ten years. Table 10.3 shows this rapid increase in the rate of consumption over the twenty-first century.

The local reservoir that supplies the water can supply a maximum of 100 billion gallons of water per year. In Year 2000, the community is using just 0.1 percent of the

---

[3]  Quoted in Deffeyes (2006).

| Table 10.3 The amount of water consumed each year by a community, assuming that consumption increases by 7.2 percent/year | | |
|---|---|---|
| Year | Gallons per year consumed | Fraction of supply consumed (%) |
| 2000 | 100 million | 0.1 |
| 2010 | 200 million | 0.2 |
| 2020 | 400 million | 0.4 |
| | . . . | |
| 2070 | 12.5 billion | 12.5 |
| 2080 | 25 billion | 25 |
| 2090 | 50 billion | 50 |
| 2100 | 100 billion | 100 |
| 2110 | 200 billion | 200 |
| 2120 | 400 billion | 400 |

*Note:* The middle column shows the water consumed each year. Dividing the rate of consumption by the maximum amount that can be supplied (100 billion gallons/yr) yields the fraction of the supply that is being consumed (right-hand column).

reservoir's capacity. This is such a small amount compared to the size of the supply that the community members do not even consider the idea that they might one day run out of water.

As consumption increases over the twenty-first century, the community uses a larger and larger fraction of the available water. This is seen in Figure 10.2, which shows the time series of the fraction consumed. By 2070, the community is consuming 12.5 percent of the supply, more than a 100-fold increase since 2000 in the rate of consumption and in the fraction of the supply used.

That same year, a small group of activists claim that the world is on the verge of exhausting its water supply. Most citizens dismiss this claim out of hand: The community is using only 12.5 percent of the supply, and it took 70 years of growth to

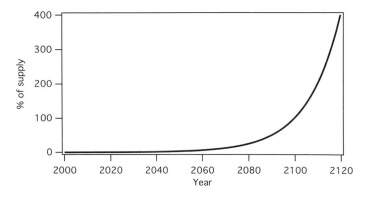

**Figure 10.2**   Consumption of water as a fraction (in percent) of total supply, based on the data in Table 10.1.

go from 0.1 percent to 12.5 percent. Based on common sense, the majority argue, it will be centuries before the water supply's limit is reached.

But the majority is wrong. As Figure 10.2 shows, around 2070 something remarkable happens: The consumption of water "turns the corner and begins heading up rapidly.[4] The reason is that, during exponential growth, the increase each doubling period is equal to the increase during all of the previous doubling periods, combined. Between 2070 and 2080, consumption will grow from 12.5 billion gallons per year to 25 billion gallons per year. So the increase in consumption between 2070 and 2080 of 12.5 billion gallons per year is equal to the total increase between 2000 and 2070. And it will only take two more doublings before consumption is 100 billion gallons per year, or 100 percent of the supply.

Now let us assume that the community suddenly realizes that it is going to reach its limits of consumption in a few decades. Through Herculean efforts to develop and deploy new technology, the reservoir's capacity is increased by a factor of four – so it begins supplying 400 billion gallons of water per year.

That is an enormous increase – but it buys far less time than you might think. If the growth rate remains constant, the community will be consuming 100 billion gallons per year in 2100, 200 billion gallons per year in 2110, and 400 billion gallons per year in 2120. This means that the new water supply limit is reached in just two doublings or twenty years. The lesson here is that, when things are growing exponentially, resource limits may be closer than you think.

In reality, of course, economic factors come into play that will reduce the growth rate. As limits to the resource are approached, the price should go up, giving consumers a clear economic signal to change their behavior. This will also encourage conservation, as well as technical development of new sources and of substitutes.

Another good and recent example is the subprime housing crisis that hit the United States in 2007. Beginning around 2003, house prices began rapidly increasing, with price increases in many metropolitan areas of 20 to 30 percent/year. Some regions, such as Las Vegas, saw even more rapid price increases (Figure 10.3). These growth rates correspond to doubling times of a few years. At that rate, a $250,000 house would become a $2 million house in about a decade (3 doubling periods). Common sense tells us that is impossible – at that rate, houses would rapidly become unaffordable. Such growth in house prices had to stop, and soon.

Although this conclusion may seem obvious in retrospect, at the time people seemed to believe this exponential growth could go on forever. Homebuyers were willing to pay ever-increasing prices and take out loans they could not possibly afford because they believed that the value of their house would always continue to appreciate. If they got in trouble, they could always sell the house at a profit. Investors believed the same thing, so they were willing to fund increasingly absurd mortgages.

Eventually, buyers with these insanely large mortgages began defaulting on them, leading to a sudden, rapid collapse of the housing market. Just a few years after

---

[4] If the exponential were infinite, the choice of where the exponential turns the corner is arbitrary and can be selected by carefully choosing the scale. However, when it is exponentially growing consumption of a fixed resource, the region where the curve turns the corner is real and set by the maximum value of the resource.

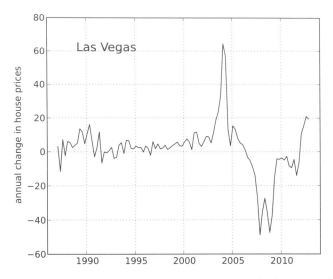

**Figure 10.3** Annual change in house prices in Nevada (data obtained from the St. Louis Federal Reserve Web site; see research.stlouisfed.org/fred2).

houses were appreciating at 20 to 30 percent/year, their value was declining just as fast. Figure 10.3 shows how quickly this transition occurred. This is a frequent occurrence when things are growing exponentially.

Probably the best-known warning of the dangers of exponential growth came about 200 years ago from *Thomas Malthus.* Malthus was an English cleric and scholar who argued that population grows exponentially whereas food production grows in a linear fashion. Linear growth means that the increase in each year is a fixed amount: Food production after $n$ years is $F(n) = an + b$, where $a$ and $b$ are constants and $n$ is the number of years. This is quite different from exponential growth, in which the increase is a fixed fraction of the growing quantity. Mathematically, it is easy to show that exponential growth will always eventually outpace linear growth, and this led Malthus to conclude that an exponentially increasing population would eventually outstrip the world's ability to feed those people, resulting in widespread starvation – what we now call a Malthusian catastrophe.

Malthus was correct that population grows exponentially. However, in the two centuries since Malthus' prediction, technological developments (e.g., development of fertilizers and pesticides) have allowed food production to increase exponentially along with population. As a result, we have not experienced this Malthusian catastrophe. We are not, however, out of the woods yet. One of the main points of this chapter is that exponential growth cannot continue indefinitely: Both food production and population will eventually cease to grow exponentially. The question is which one plateaus first. If exponential food production growth stops before exponential population growth does, then Malthus's prediction may yet come true. If, however, population growth ceases before food production growth, then Malthus may be forever wrong. Time will tell.

# 10.4 Discounting

In this section, I describe the financial concept of discounting, which plays a key role in evaluating policy options for dealing with climate change.

## 10.4.1 The time value of money

Suppose you know today that you will incur an expense of $25,000 in fifteen years. Such eventualities occur frequently in the world of corporate finance and other business areas. How much would you be willing to pay *today* to eliminate that future expense? One way to answer this is to determine how much you would have to invest today in order to have $25,000 in fifteen years. We can get the answer by rearranging Equation 10.1:

$$P = F/(1 + r/100)^n \qquad (10.4)$$

Here $F$ is the expense, which will be incurred in $n$ years, $r$ is the interest rate in percent, and $P$ is the amount you need to invest today. Given an interest rate of 5 percent, we need to invest about $12,000 today in order to have $25,000 in fifteen years. In other words, we can view $12,000 today as being equal to $25,000 in fifteen years.

One conclusion you can draw from this is as follows: *Money in the future is worth less to you than money today.* That general conclusion is probably obvious to most people, but this calculation allows us to answer the question of how much less. The parameter $r$ in these projects quantifies the rate that money loses value as it recedes into the future – each year, money loses $r$ percent of its value. In these types of problems, $r$ is frequently referred to as the *discount rate,* and the value today of a future expense or benefit is referred to as the *present value.* The process of calculating the present value of a future cost or benefit is referred to as *discounting*.

### An example: What would you do?

You can use discounting to help make financial decisions. Imagine you walk into an electronics store, searching for a new television. You select one and are informed that you have two payment options: You can get it "with no money down" and pay $1,100 in one year, or you can pay $1,000 today. Which option do you choose?

Note that $1,100 in one year has a present value of $1,100/(1 + r/100)$, where $r$ is the discount rate. If you choose a discount rate of, say, 5 percent, then the present value is $1,047. This is more than $1,000, meaning that $1,100 in one year is more expensive than $1,000 today. You want to pay as little as possible for the television, so you therefore choose to pay $1,000 today.

For a discount rate of 15 percent, the present value of $1,100 in one year is $956, so in that case $1,000 today is more expensive than $1,100 in one year – and you would therefore prefer to pay $1,100 in one year. If you choose a discount rate equal to

10 percent, then the present values are equal, and you would have no preference about paying $1,000 today or $1,100 in one year.

In the policy debate over climate change, our choice is between spending money today to reduce emissions of greenhouse gases, thereby reducing the impacts of climate change in 50 to 100 years, or doing nothing now and spending more money dealing with the impacts of climate change in a few decades. So, for example, our choice might be to spend $100 billion today or $1 trillion in 100 years.

Discounting allows us to quantitatively compare these two options. If we assume a discount rate of 3 percent, then the present value of $1 trillion in 100 years is $10^{12}/(1.03^{100}) = \$52$ billion. This is less than the alternative of spending $100 billion today, so (from a purely financial perspective) we would prefer to pay $1 trillion in 100 years than $100 billion today. This type of analysis, in which you compare the present value of the costs and benefits of various options, is a *cost-benefit analysis,* and economists frequently use it to provide guidance to policymakers about alternative policy options.

## 10.4.2 The discount rate

In both the television purchase and climate examples just given, the answer we get is strongly dependent on the discount rate. In the climate example, a discount rate of 3 percent yields the conclusion that we would prefer to do nothing now and pay later to address the impacts of climate change. However, if the discount rate were 2 percent, then the present value of $1 trillion in 100 years is $138 billion, and we would rather pay $100 billion today to reduce emissions.

So how do we determine the correct discount rate? The discount rate really is a combination of two different judgments. First is what is known as *time discounting,* which is the preference to consume now rather than later. If offered $100 now or $100 in one week, just about everyone would choose to get the money now. After all, why would you wait? Experiments show that animals also exhibit this behavior. I know that if I give my dogs the choice of having dinner now or in an hour, they tell me in no uncertain terms that they want to eat now. In other words, most people (and dogs) have a positive time discount rate: Goods and services now are worth more than the same goods and services in the future.

The climate problem covers periods longer than a human lifetime. In that case, the time discount does not represent our preference for us to consume now rather than later. Rather, it represents our preference for us to consume rather than future generations. Given that consumption can be roughly equated to welfare, the time discount rate then expresses how much we value our welfare above the welfare of future generations.

From a moral standpoint, most people agree that it is unethical to place a higher value on our own welfare over that of future generations, which implies that the time discount rate should be set to near zero. Nonetheless, it is clear from our society's actions, such as our low rate of savings and our failure to address big problems facing future generations (e.g., climate change, budget deficits), that we do indeed value our own generation more highly, which implies a positive time discount rate.

The other part of the discount rate is known as *growth discounting,* and it reflects the fact that a dollar means more to poor people than it does to rich people. For example, if a billionaire is walking down the hall and sees a $1 bill on the floor, would he stop to pick it up? Probably not – if you have 1 billion dollars, another dollar does nothing to improve your welfare. If you are living in poverty, however, you are most certainly going to stop and pick up the $1 bill – it might be the difference between having dinner that night or not.

In economics jargon, the utility of $1 to the billionaire is much lower than the utility of $1 to the person living in poverty. In the case of climate change, we expect future generations to be richer than we are, just like we are richer than those living 100 years ago. And because future generations are richer, they will be better able to pay costs associated with climate change than we are. This suggests a preference for our generation to push the costs of addressing climate change onto richer, future generations. This preference is expressed in the growth discount rate, which is the rate at which the utility of money – how much each dollar means to society – declines with time as the world gets richer.

The discount rate used in present-value calculations is determined by combining the time and growth discount rates. Unfortunately, the choice of both time and growth discount rates is as much of a value judgment as an objective fact, and there are wide disagreements about what discount rate to use. Some economists argue the discount rate should be near zero, whereas others argue for higher values such as 4 percent.

This makes a huge difference in climate problems. For future impacts of $1 trillion in 100 years, the different discount rates yield present values of $1 trillion (for a 0 percent discount rate) or $19 billion (for a 4 percent discount rate). The policy implications of these hugely different estimates of the present value of climate impacts are completely different.

## 10.5 Putting it together: The social cost of carbon

Imagine that you emit a ton of carbon dioxide to the atmosphere. As discussed in the first half of the book, this ton of carbon dioxide will warm the climate over many thousands of years – until the carbon cycle completely removes it from the atmosphere. As described in Chapter 9, this warming imposes costs on our society. For example, increases in sea level require construction of expensive defenses against the sea (e.g., sea walls) or relocation of communities being inundated. And this is just one cost from carbon dioxide; there are also costs associated with rising temperatures, changing precipitation patterns, ocean acidification, etc. Calculating the total cost imposed on society by this ton of carbon dioxide requires adding up all of the costs from all of the impacts.

Once you know the total cost of this ton each year, the costs can be discounted back to today. As an example, imagine that the ton causes $1 of damage every year for 300 years. Assuming a discount rate of 3 percent, the present value of the cost of impacts during the first year is $1/1.03$, the second-year's damages is $1/1.03^2$, and year n's damages is $1/1.03^n$.

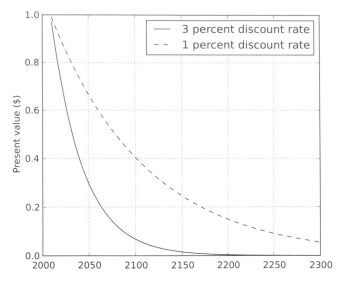

**Figure 10.4**    The present value of $1 of climate damages, as a function of the year the damages took place. The calculation is shown for two discount rates.

Figure 10.4 shows the present value of $1 of climate damages for two discount rates, as a function of when the damages occur. The figure shows that, as the damages recede into the future, the present value of the damages declines rapidly. For the higher discount rate, costs beyond about 2200 (200 years from now) have a present value of zero. For the lower discount rate, the present value drops off more slowly and a small fraction of the damages in 2300 are contributing to the present value.

We can calculate the total cost to us today from the emission of this ton of carbon dioxide by summing the discounted costs for every year. This total cost is frequently referred to as the *social cost of carbon*, and in this simple example it equals $33 and $94 for the discount rates of 3 percent and 1 percent, respectively.

Sophisticated expert estimates of this quantity from a recent assessment are listed in Table 10.4. There are several things worth noting in this table. First, seemingly small changes in the discount rate lead to large changes in the social cost of carbon. This emphasizes the central role of the discount rate in the climate-change policy debate.

Second, the social cost of carbon rises throughout the twenty-first century. The reason for this is that damage from climate change is non-linear: each degree of warming produces more damage than the previous degree. Thus, as we add carbon dioxide to the atmosphere, the cost of the damage from each additional ton is more than the cost from the previous ton.

Third, comparing the 3% column and the 3%, 95th percentile column gives some idea of how wide the range of estimates are for a single discount rate. The 3% column is the average of all of the estimates, while the 3%, 95th percentile column is the value exceeded by 5 percent of the estimates using a 3 percent discount rate. The large difference between these values says that, even at a single discount rate, there is

**Table 10.4** Estimates of the social cost of carbon (in 2007 dollars) for different discount rates. The 5%, 3%, and 2.5% columns are the averages of the estimates from a large number of economic models; the 3%, 95th percentile column is the value exceeded by 5% of the estimates of the estimates using a discount rate of 3%.

| Year | Discount rate | | | |
|------|-----|-----|------|--------------------|
|      | 5%  | 3%  | 2.5% | 3%, 95th percentile |
| 2010 | $40 | $121 | $191 | $330 |
| 2015 | 44  | 139 | 213  | 400  |
| 2020 | 44  | 158 | 239  | 473  |
| 2025 | 51  | 176 | 257  | 528  |
| 2030 | 59  | 191 | 279  | 584  |
| 2040 | 77  | 228 | 319  | 705  |
| 2050 | 99  | 261 | 360  | 811  |

*Note:* This table is the value per ton of carbon; the value per ton of carbon of carbon dioxide is 1/3.67 of this value.

*Source:* Technical Support Document: Technical Update of the Social Cost of Carbon for Regulatory Impact Analysis, Under Executive Order 12866 (www.whitehouse.gov/sites/default/files/omb/inforeg/social_cost_of_carbon_for_ria_2013_update.pdf). This is an update of a 2010 analysis (www.epa.gov/otaq/climate/regulations/scc-tsd.pdf).

a wide range of estimates of the social cost of carbon, with some models estimating much larger values than others.

The wide range of estimates in Table 10.4 underscores two primary difficulties in estimating this quantity. First, the value you get is sensitive to the choice of discount rate, and there is little agreement among economists as to what the right value is. Second, putting a dollar price on the impacts of climate change requires a set of linked predictions, all of which are highly uncertain. We need to predict how the climate will change in individual regions, how that climate change will affect people and ecosystems, and how people and ecosystems will in turn respond to these climate changes. Then, a dollar value must be assigned to the impacts.

For some impacts, a dollar value is easy to assign. For goods and services sold on the open market (e.g., agricultural products, farmland, and coastal property), the market value can be used as an estimate of their social value. Other impacts are much more difficult to value. A good example is the existence of polar bears. They have little market value, but (to me, at least) they do have social value – I believe that driving them to extinction would be a great loss to humanity. Most people would agree that there is some value in their continued existence, but there might be great disagreement over exactly how much.

The difficulty in calculating the overall costs of climate impacts is responsible for the large spread of estimates in Table 10.4. However, despite the large range of estimates, there are a few things we can conclude with confidence:

1) The social cost of carbon is not zero. Every credible economic analysis of climate change found has found that there are costs from the emissions of greenhouse gases and the associated climate change.

2) While there is uncertainty in our estimates of the social cost of carbon, very large costs cannot be ruled out. Such large costs are associated with dire climate change impacts, and the possibility of this bleak future drives much of the concern about the climate problem.

Despite the uncertainty in this quantity, regulators must nonetheless assign a single value for the social cost of carbon in order to use it in cost-benefit analyses and other regulatory decisions. In 2013, the Obama Administration selected $132 per ton of carbon ($36 per ton of carbon dioxide) as our best estimate of today's social cost of carbon.

## 10.6 Chapter summary

- Exponential growth means that the rate of growth is directly proportional to the present size. Anything growing at "$x$ percent/year" is growing exponentially.
- A quantity $P$ growing at $r$ percent/year will grow to $P(1 + r/100)^n$ after $n$ years. The time for a quantity growing exponentially to double is frequently referred to as the doubling time; it is approximately equal to $72/r$, and this shortcut is known as the "rule of 72."
- Exponential growth tends to be end loaded, meaning most of the growth occurs at the end: 50 percent of the growth occurs during the last doubling period, and 97 percent of the growth occurs during the last five doubling periods.
- Exponential growth cannot go on forever, and when it ends, it often ends abruptly.
- Discounting refers to the process of calculating the present value of some future expense or benefit. Such calculations require a discount rate, which is the rate at which money loses value in the future. In general, money in the future is worth less than money today.
- Cost-benefit analyses compare the present values of the costs and benefits of a range of policy options. The best option (from a financial point of view) is the one with the largest present-value net benefit.
- The discount rate is determined by two judgments: the time discount rate, which is our preference for consuming now rather than later, and the growth discount rate, which reflects the fact that future generations are expected to be richer so they can pay a bigger share of the costs. There are vigorous disagreements over what discount rate we should use.
- The social cost of carbon is the cost, discounted to today, of the future impacts due to the emission of one ton of carbon dioxide to the atmosphere.
- The social cost of carbon is a highly uncertain quantity, due to uncertainty in the discount rate and from difficulty in assessing the monetary value of climate change impacts. Despite the uncertainty, we can be confident that there are net costs to climate change, and these costs could potentially be very large. In 2013, the Obama Administration selected a value of $132 per ton of carbon as our best estimate of the social cost of carbon.

# Additional reading

Much of the material in this chapter was motivated by several YouTube videos of lectures by Dr. Albert Bartlett on exponential growth. I highly recommend watching them. Dr. Bartlett also wrote numerous papers on exponential growth, many of which are collected in his book, *The Essential Exponential!* See www.andrewdessler.com/chapter10 for links to Dr. Bartlett's material as well as other related resources.

# Terms

Cost-benefit analysis
Discounting
Discount rate
Doubling time
Exponential growth
Growth discounting
Interest rate
Malthus
Present value
Rule of 72
Social cost of carbon
Time discounting

# Problems

1. You invest $1 at a 10 percent interest rate for fifty years.
    a) Use Equation 10.1 to calculate how much you have after fifty years.
    b) How many doubling periods does the investment experience?
    c) Use Equation 10.2 to calculate how much you have after fifty years.
2. You invest $50 at a 7 percent interest rate for thirty years.
    a) Use Equation 10.1 to calculate how much you have after thirty years.
    b) How many doubling periods does the investment experience?
    c) Use Equation 10.2 to calculate how much you have after thirty years.
3. a) How many doubling periods do you have to wait for 1 cent to grow to $100 trillion? (Calculate to the nearest integer.)
    b) At an interest rate of 7 percent, about how long does it take for that many doublings to occur?
4. Would you rather pay $1 trillion dollars of damages from and adaptation to climate damage in fifty years or pay $50 billion dollars today to reduce emissions and avoid the climate change? Use discount rates of 0 percent, 2 percent, 4 percent, 6 percent, and 8 percent.

5. You go into a big-box electronics store to buy a flat-screen television. You have two options: pay $1,400 today or $1,450 in one year. Which do you choose? You have to estimate a discount rate to do this. How did you choose your discount rate?

6. Lotteries often give you the option of taking a lump-sum payment now or a fixed amount every year for, say, twenty-five years. For this question, assume that the lump-sum payment is $3 million and the yearly payments are $250,000 each year for twenty-five years. The first payment is made immediately, so it is not discounted, and subsequent payments are made every year thereafter.

   a) Find the discount rate where the present value of the twenty-five-year cash stream is equal to the lump-sum payment.

   b) If your discount rate is higher than this, should you take the lump-sum payment or the period payments?

7. In the National Football League draft, a pick in this year's draft is worth a pick in a lower round in a future draft (e.g., you might trade a second-round pick in this year's draft for a first-round pick in next year's draft). Explain how this is consistent with the concept of discounting.

8. a) Imagine you have a dollar bill. If you double it, you have two bills. If you double again, you have four bills. If you double again, you have eight bills, and so on. Given that a bill is 0.1 mm thick, how many doublings do you have to go through before you have a stack that reaches from the Earth to the moon? (The moon is 360,000 km away.)

   b) How many doublings do you need to get a stack that goes halfway to the moon?

   c) How many doublings to you need to get 1 percent of the way to the moon?

9. a) Consider the choice between paying $10 million today to reduce emissions that cause climate change or $1 billion in 100 years to adapt to a changing climate. What would the discount rate have to be in order for these two choices to be equal?

   b) Using that same discount rate, what would be your preference if the expense was in 50 years instead of 100?

10. You are inside the Houston Astrodome, in the rafters just below the roof, 160 ft above the field. A wizard puts a tiny magic drop of water on the pitcher's mound, and the drop starts doubling in volume every minute. After 100 minutes, there is 5 ft of water on the field, and the depth continues to double every minute. How many minutes do you have before the water reaches you?

11. Calculating the cost of climate change.

    a) If the economy grows at 3 percent/year, how many times larger than today will it be after 100 years?

    b) Imagine that addressing climate change reduces economic growth from 3 percent to 2.9 percent over the century. How much smaller is our GDP in 100 years?

    c) How many additional years of growth at 2.9 percent need to occur until the GDP is as large as 100 years of growth at 3 percent?

    d) Put yourself in the shoes of a future citizen: Given how much richer people will be in 100 years (that is the answer to part a), should we be concerned about the loss of wealth due to a reduction in growth from 3 percent to 2.9 percent that we calculated in part b?

12. Imagine that you can save the polar bears if you pay a fee every year. If you do not pay the fee, they go extinct. How much would you pay each year to keep polar bears alive?

# Fundamentals of climate change policy

In the previous chapters of this book, we have seen that 1) the Earth is warming, 2) most of the recent warming is very likely due to human activities, 3) warming over the next century will likely be a few degrees Celsius, and 4) such warming carries with it a risk of serious, perhaps even catastrophic impacts for humans and the planet's ecosystems.

Given those facts, what shall we do about climate change? Science, it turns out, is just one of several factors needed to answer this question. Deciding what to do also requires information about the options available to us to respond to climate change, and the costs, benefits, and risks of each option. In addition, we must consider not just monetary costs but also the moral implications of each policy. In this chapter, I will outline the various options available to us to address climate change.

Our responses to climate change can be broadly split into three categories: adaptation, mitigation, and geoengineering. *Adaptation* means responding to the negative impacts of climate change. If climate change causes sea-level rise, an adaptive response to this impact would be to build seawalls or relocate communities away from the encroaching sea. *Mitigation* refers to policies that avoid climate change in the first place, thereby preventing impacts such as sea-level rise from occurring. This is accomplished by reducing emissions of greenhouse gases, usually through policies that encourage the transition from fossil fuels to energy sources that do not emit greenhouse gases.

*Geoengineering* refers to active manipulation of the climate system. Under this approach, our society would continue adding greenhouse gases to the atmosphere, but we would intentionally change some other aspect of the climate system in order to cancel the warming effects of the greenhouse gases. For example, we could engineer an increase in the albedo of the Earth. If done correctly, this could stabilize the global-average climate despite continuing emissions of greenhouse gases. In the rest of this chapter, we explore each of these options in detail.

## 11.1 Adaptation

As temperature, precipitation, sea level, and other components of our climate change, we can adapt our way of life to adjust to these changes. There are several advantages to relying on adaptation as the main response to climate change. First, because many of the worst impacts of climate change will occur in the second half of the twenty-first century, adaptation allows us to wait for decades before we must start adapting. There are clear advantages to doing this – for example, it allows us to resolve uncertainty on how climate will change and focus our efforts on the severe impacts, without wasting any effort dealing with impacts that do not materialize.

Second, if the past is any guide, we can expect future generations to be richer than we are and better able to bear the costs of adaptation (this is the factor that growth discounting, discussed in the last chapter, accounts for). Third, many of the adaptations necessary to address climate change will simultaneously benefit society in other ways. Decreasing a community's vulnerability to sea-level rise caused by climate change will also decrease the vulnerability to extreme sea-level events caused by hurricanes and other severe storms. Conserving water to address decreased freshwater availability will also decrease the community's vulnerability to droughts or to increased demand caused by population growth. An improved public health infrastructure designed to head off disease outbreaks in a warmer world would also help decrease a society's vulnerability to pandemic flu and other nonclimate public health issues.

Finally, of all of the possible responses to climate change, adaptation requires the least intrusion of government into the private lives of individual citizens. In fact, strictly speaking, adaptation requires no government intervention at all. If sea level rises and the government does nothing, individual citizens will not just sit there and be submerged, they will make individual decisions about how to respond: whether to build coastal defenses, move to higher ground, etc. If climate changes in agricultural areas, affected farmers can decide whether to relocate, change farming practices, etc. Because of this, those philosophically opposed to government regulation find adaptation to be a less serious infringement on personal liberty than mitigation or geoengineering.

However, some of these advantages are largely illusory. Waiting until the impacts of climate change are obvious is much more expensive than adapting in advance. For example, if you are building an airport on the coast that may last 100 years, it is better to spend money today to build sea-level rise into the design than it is to wait for the airport to flood in 70 years and then face much higher costs of fixing the damage from the flood as well as making the infrastructure less vulnerable to future floods (e.g., sea walls, levees).

And the idea that adaptation can be a wholly local response is also largely illusory because adaptation takes huge resources. For example, building a sea wall to protect a community against rising sea levels is expensive, and most individuals and even most individual communities, simply cannot afford to build one. Because of this, it is generally agreed that the significant adaptation efforts require national governments or international institutions to provide resources. And such national or international assistance to a local community often makes sense. If sea-level rise submerges Miami, and the resulting economic disruption hurts the entire U.S. economy, then the U.S. federal government might be justified in paying for seawalls to prevent that from happening. Or if climate impacts in China threaten to destabilize the world economy, international assistance to help the Chinese deal with the impacts may be appropriate.

In addition to direct aid, governments can also implement regulations to encourage citizens to adapt to a changing climate. Regulations promoting water conservation, for example, would help communities adapt to decreased freshwater availability. Governments can also eliminate existing regulations that encourage us to be poorly adapted to the present climate and that increase our vulnerability to climate change. A good example is flood insurance. People love to build houses near bodies of water, such as the ocean. However, the downside of this is that flooding may occasionally destroy their houses. Without flood insurance, many people would find it too risky to build in flood-prone areas because they could not afford to have their houses destroyed. With flood insurance, however, people can afford to live in flood-prone areas; if their houses are destroyed by flood, the insurance covers the loss. In this way, flood insurance actually encourages people to build where it is going to flood.[1]

A third way government policy can facilitate adaptation is by providing reliable information about climate change, as well as possible responses, including technical assistance. Such information would help people plan for climate change and adapt in advance rather than waiting for disaster to strike and then dealing with climate change.

There are also significant disadvantages to relying on adaptation as the main response to climate change. Because adaptation to climate change requires resources, the effects of environmental disruption are felt most strongly by the poorest and most vulnerable in any society. This was ably demonstrated when Hurricane Katrina hit New Orleans in 2005. When the storm hit, the wealthier residents of New Orleans, those with resources such as credit cards and automobiles, simply left town. The poorest residents of the city, who lacked resources to evacuate, were stranded in New Orleans and made up the vast majority of those killed. The poor also tended to live in more poorly constructed housing, frequently located in marginal areas, such as swampy or low-lying, flood-prone land.

Once the storm passed, the wealthier residents of New Orleans had the resources to reconstruct their lives – either to return and rebuild or to start anew somewhere else. The poorer residents had to rely on government assistance to rebuild their lives. Because the United States is rich, it was able to provide assistance such as money and temporary housing that kept Katrina from being the much larger humanitarian disaster it would have been for a poorer country.

We also see connections between vulnerability and wealth in two recent earthquakes. In early 2010, a magnitude-7.0 earthquake ravaged Haiti – it killed more than 200,000 people, made 1 million people homeless, and heavily damaged much of the country's built structures. A few weeks later, a magnitude-8.8 earthquake hit Chile. Although this earthquake was about 500 times stronger, it killed just a few hundred people. Much of the difference in death toll can be attributed to Chile's greater wealth. Richer countries have the luxury of spending more money on infrastructure, so buildings in Chile were built to more stringent standards and most did not collapse during the quake.

The connection between the severity of climate change impacts and wealth means that adaptation fails a fundamental fairness test. The world's rich economies have built their wealth by consuming massive amounts of energy, which means that these economies are responsible for most of the global warming over the past two centuries. Yet their very richness allows these countries to deal most effectively with the impacts.

---

[1] By encouraging the undesirable outcome, this is an example of what is often referred to as a "moral hazard."

The poorest countries in the world are responsible for very little of the greenhouse gases in our atmosphere today, yet they are least capable of dealing with the impacts.

Because of this, an adaptation-only response is frequently viewed as morally problematic because it abandons the poorest people in the world to the impacts of climate change that they did not cause. To the extent that adaptation is necessary, there is general agreement in the international community that rich countries must help the poorer countries of the world adapt to the impacts of climate change.

Although the rich of the world have many more resources than the poor to deploy against the impacts of climate change, do they have enough? That depends on your definition of "successful" adaptation. If you define successful adaptation as minimal survival of the human species, then humans can successfully adapt to almost any climate change. After all, we are a resilient species and it is hard to believe that any climate change will lead to our extinction.

A more realistic standard of success is maintenance of our standard of living. The challenge with this metric is that there is no single agreed-upon way to measure it. At its most restrictive, the standard of living can be equated to the amount of goods and services consumed (e.g., GDP per person). A more expansive definition might also include the value of nonmarket activities such as the opportunity to hike in the same pristine wilderness you did as a child, the satisfaction of seeing the leaves in New England change colors every fall, or appreciation for the existence of polar bears, even if you never actually see one. None of these are fully accounted for in GDP.

Whereas GDP is relatively easy to measure, estimating the value of nonmarket activities is much more difficult. Most of us agree that a world with polar bears is more valuable than a world without, but exactly how much more? Because there is no universal answer to this question, a particular strategy for adaptation might be viewed as a success by someone who puts less value on polar bears and a failure by someone who puts more. This can lead to starkly different views about our ability to adapt to climate change.

Despite this, there are a few general rules about successful adaptation that we can have high confidence in. First, the more warming we have, the harder it will be to successfully adapt – regardless of our definition. Second, for a given amount of climate change, how we define success in adaptation plays a key role in determining our ability to successfully adapt. The more expansive the definition, that is, the more value we place on maintaining the environment in its present state, the harder it will be to successfully adapt. In fact, for those who put the highest value on maintaining the present environment, climate change already experienced combined with committed warming over the next few decades may be so large that it is no longer possible for us to successfully adapt.

**The bottom line on adaptation:** Because of lags in the climate system, as well as lags in the economy, some future climate change is unavoidable. To the extent that this warming cannot be stopped, we must adapt to it. Thus, adaptation must be a part of our response. However, relying *entirely* on adaptation as our response is problematic. Adaptation requires resources, and many of the world's poorest inhabitants have few resources and therefore little ability to adapt. As a result, even low-to-moderate climate change will impose harsh impacts on these people. Many view this as morally unacceptable because the world's poor have contributed little to the problem of climate change.

Rich countries have more resources to address the problem, and for low-to-moderate climate change they may well be able to successfully adapt. But if climate change is at the upper end of the range of predictions discussed in Chapter 8, even rich countries may not have enough resources to adapt. Thus, relying entirely on adaptation as our only response to climate change is a titanic gamble for the rich that climate change over the next century will not be severe. For the poor, it is no gamble at all – climate change will impose significant hardships. All of this, of course, depends to some extent on the definition of successful adaptation.

As a result, adaptation-only policies are not seriously considered in the climate policy debate, and there is wide agreement that mitigation must be part of our solution to the problem of climate change.

# 11.2 Mitigation

Mitigation refers to reductions in emissions of carbon dioxide and other greenhouse gases, thereby avoiding the impacts of climate change by preventing the climate from changing in the first place. Because relying entirely on adaptation is a risky strategy, most policymakers view mitigation as the centerpiece of any long-term climate change policy. There are several approaches that could be used to reduce emissions, and I discuss the range of options available in this section.

Before we get into the details of mitigation, it is important to make clear the size of the required reductions. As we will discuss in Chapters 13 and 14, almost every country in the world has agreed with the judgment that warming of more than $2\,^{\circ}$C above preindustrial temperatures would be considered dangerous. Stabilizing at or below this temperature would require a reduction in the global emissions by the middle of the twenty-first century of greenhouse gases by 50 to 80 percent below today's emissions levels and reaching near-zero emissions later in the century. This is a challenging but achievable target, and we will discuss it in more detail in Chapter 14.

In Chapter 8, we explored the factors that control emissions of greenhouse gases: population, affluence, and technology. Thus, we can recast the problem of reducing emissions into the problem of reducing one or more of these factors until emissions reach a desired value. The first factor is the world's population. With fewer people on the planet consuming goods and services, emissions would certainly decrease. Some societies have already implemented policies to actively influence the size of their population. China, for example, adopted a "one-child policy," which limits the number of children a family can have to one, although there are many exemptions. This policy has significantly reduced China's population growth rate, although the total population is still increasing.

Reducing emissions through population control would require more than just a reduction in the rate of population growth – it would require a significant reduction in the actual number of people on the planet. Such an effort would conflict with deeply held religious, social, and cultural traditions surrounding reproduction and family size in many countries. It also creates other demographic and societal problems – as China has discovered in response to its one-child policy. As a result, efforts to combat climate change by using policies explicitly targeted at reducing the Earth's population are viewed as politically unachievable, and there are no serious discussions of this approach.

A second option is to reduce the world's consumption of goods and services. If each person consumed less, the amount of energy consumed, and therefore emissions, would decrease. Like population, solving the climate problem through consumption would require not just stopping growth of consumption but deep reductions in it. There are several problems with solving climate change this way. First is a political problem – people equate consumption with well being. That is why all politicians strive for increased consumption (which they call economic growth), and no politician who wants to keep his job would agree to a policy that steeply reduces it. Thus, reducing consumption is something that most countries simply will not agree to.

Then there is the two billion or so of the world's poorest inhabitants who live in extreme poverty, with incomes of a few dollars per day. These people do not have the basic necessities of life – food, clean water, shelter – and lifting them out of poverty requires economic growth, which translates into increasing consumption. Efforts to limit consumption to address climate change might mean preventing these people from escaping poverty. Such an outcome is generally viewed as morally unacceptable: We cannot solve climate change on the backs of the world's poorest people.

Thus, like population control, there are no serious efforts to address climate change by reducing the world's level of consumption. The rich world does not want to do it, and it is ethically problematic to impose such a policy on the world's poor.

If neither population nor consumption has any chance of being reduced, then by process of elimination it is the technology term, also referred to as the greenhouse-gas intensity, which must be reduced in order to reduce emissions. As I discussed in Chapter 8, the technology term is a measure of how much greenhouse gas is emitted per dollar of GDP, which itself can be broken into two constituent terms: the energy intensity, a measure of how much energy it takes to generate \$1 of GDP (J/\$), and the carbon intensity, a measure of how much greenhouse gas is emitted to generate a Joule of energy ($CO_2$/J). If your memory of this is hazy, you may want to review Chapter 8.

Reducing greenhouse-gas intensity therefore requires reducing energy intensity, carbon intensity, or both. Energy intensity is determined to a large extent by the efficiency with which the economy uses energy. Today's society wastes a tremendous amount of energy, and improving our energy efficiency would not only reduce our emissions of carbon dioxide but would have many co-benefits, such as saving us money and reducing air pollution. Because of the co-benefits, energy-efficiency improvements would make sense even if climate change were not a problem.

Can efficiency improvements lead to large enough reductions? To reduce emissions by 50 to 80 percent over the next few decades, which is about what is required to stabilize the climate with less than 2 °C of warming, would require reducing emissions by approximately 2 percent/year. If the world's total GDP (the product of population and affluence) grows by 3 percent/year, then energy intensity would need to decline by 5 percent/year or so to achieve the necessary reductions in emissions.

Historically, energy intensity has decreased at roughly 1 percent/year. This can likely be maintained, and some improvement may be possible, but most experts do not believe that rates of decline of energy intensity of 5 percent/year for several decades are realistic. So although improvements in energy efficiency can contribute to emissions reductions and are something we should be doing now, they are likely only going to play a supporting role in solving the climate problem. We therefore conclude that it is reductions in the carbon intensity term, the amount of carbon dioxide emitted per Joule of energy generated, that are required to stabilize the climate.

## 11.2.1 Technologies to reduce carbon intensity

Reducing carbon intensity is code for switching from conventional combustion of fossil fuels to energy sources that do not release greenhouse gases – often referred to as *carbon-free* or *climate-safe* energy sources. These include nuclear energy, carbon capture and sequestration, and energy sources known as *renewable energy,* because these energy sources are not depleted when utilized: primarily hydroelectric, solar, wind, and biomass energy.

Solar energy is one of the most frequently discussed renewable energy sources. There are actually two different ways to generate energy from sunlight: *solar photo-voltaic* or *solar thermal* methods. Photovoltaic energy is the most common form of solar energy, and you can see it in operation in the form of solar panels located on houses or buildings (and also on satellites and the Space Station). It takes advantage of the fact that, when exposed to light, certain materials such as silicon produce electricity. Solar thermal energy, in contrast, uses mirrors to concentrate sunlight on a working fluid (such as an oil, molten salt, or pressurized steam), heating it to several hundred degrees Celsius. This hot fluid boils water and generates high-pressure steam that is used to turn a generator, producing electricity.

Solar energy is in many respects the Holy Grail of renewable energy. As we calculated in Chapter 4, the amount of solar energy falling on the planet is staggering – more than 100,000 TW. This is an enormous amount of energy compared to the amount humans consume, about 15 TW. There are, however, problems with the large-scale adoption of solar energy. One is intermittency – the Sun shines only during daytime and when not obscured by clouds. Thus, solar energy may require additional mechanisms to ensure the reliable twenty-four-hour availability of power consumers expect. Another is the area required to generate solar energy. Taking into account the intermittency and other efficiency issues, solar energy can supply power at a level of approximately 10–20 $W/m^2$. To satisfy all human energy needs would therefore require roughly 1 million $km^2$ to be covered with solar energy collectors, corresponding to 0.2 percent of the Earth's surface. Although this is a large area, it is comparable to the total area covered by cities, so there is no reason to believe that it is impossible for humans to construct the number of collectors needed for that much solar energy.

Another frequently mentioned renewable energy source is wind. This is a mature technology – the Dutch have been using wind energy for hundreds of years to do useful work, such as pumping water. Today's electricity-generating windmills, often referred to as wind turbines, are quite a bit larger and more sophisticated. The largest ones are 130 m tall, the same as a forty-story building, with 125-m blades. A single one of these wind turbines can generate as much as 6 MW of power, so that a few hundred can replace a conventional fossil-fueled power plant.

Wind also has the problem of intermittency. The wind does not blow everywhere nor does it blow all the time, so, on average, the 6-MW wind turbine actually produces roughly one-third of its maximum value. Taking into account the intermittency of wind, as well as the fact that windmills must be spaced apart so that they do not interfere with each other, we find that a wind farm generates power at a level of approximately 2 $W/m^2$. Therefore, to satisfy human energy requirements would require covering approximately 1.5 percent of the Earth's surface area with wind farms containing a few million windmills. It should be noted that putting up windmills does not preclude using the land simultaneously for other activities, such as agriculture.

Although it would undoubtedly be an enormous undertaking to construct the number of solar collectors and windmills needed to produce that much solar or wind power, human industry has produced amazing feats in the past. During World War II, for example, the countries of the world produced hundreds of thousands of airplanes, tens of thousands of tanks, thousands of ships, and massive amounts of other equipment of war. Compared to that, putting up a few million windmills does not sound so hard.

Wind and solar energy have been growing rapidly over the recent past and are emerging as important contributors to our energy supply. However, these sources remain (generally) more expensive than electricity from fossil fuels, and the intermittency problem is still being worked out. Thus, we are not yet on the verge of a wholesale transition of our energy supply to these renewable sources.

Biomass energy is another renewable option; it refers to the process of growing crops and then burning them to yield energy. Because the carbon dioxide released from burning biomass was absorbed from the atmosphere during the growth of the plant, there is no net increase in carbon dioxide in the atmosphere. It is an intuitively attractive energy source, but there are several issues that must be considered. First, the rate of photosynthesis limits the power generated by biomass to roughly 0.6 $W/m^2$ of farmed land. Thus, to generate 15 TW would require that 15 percent or so of the land surface be devoted to growing biomass for energy – comparable to the area presently under cultivation today.

The enormous land requirement is problematic. We know from experience that much of the additional land will come from clearing forest. This deforestation releases carbon dioxide into the atmosphere, and it causes a host of other local environmental impacts, such as loss of native biodiversity and ecosystem degradation. The second problem is that the farming methods used to grow the biomass have to be carefully considered. Production of fertilizer, for example, requires large inputs of energy, mainly from fossil fuels. If fertilizer is used in the growth of the biomass, it might take as much fossil fuel energy to grow the biomass as is saved by burning the biomass to produce energy.

Finally, it is becoming clear that using food, such as corn, as feedstock for biomass energy severely stresses the food supply. The increased competition for food raises food prices, an impact disproportionately felt by the poor. The hope is that a technological breakthrough will allow us to produce energy from waste biomass that does not have other uses, such as the waste from corn processing (e.g., corn stalks, corn cobs) or cellulosic biomass such as switch grass. Many scientists are currently working on methods to produce biomass energy from these waste sources. Despite these difficulties, biomass, particularly in the form of corn-based ethanol, already provides a few percent of U.S. motor fuels and is slated by act of Congress to become a major source of automotive fuel in the United States over the next decade.

Biomass energy systems are a promising technology, but any biomass system must be carefully constructed from end to end to ensure that carbon emissions are actually reduced (e.g., limiting how much fertilizer is used, where the land comes from). In addition, new technologies that allow biomass energy to be extracted from nonfood biomass must also be developed. Thus, large-scale biomass energy production may be further in the future than large-scale wind and solar.

Hydroelectric energy is the most widespread renewable energy source in the world today, providing 16 percent of the world's electricity. Despite the many advantages of this energy source, it seems unlikely that this power source can be greatly increased. Many of the world's big rivers are already dammed, and new dams often cause local environmental problems that generate significant local political opposition.

One of the most contentious options for reducing greenhouse-gas emissions is nuclear energy. Currently, nuclear reactors generate nearly 16 percent of the world's electricity. Although nuclear energy is not technically a renewable energy source, with the technology to recycle and reprocess spent nuclear fuel, there are centuries' worth of uranium in the ground, even assuming a massive expansion of the world's nuclear generation capacity. Nuclear is a mature technology, so there is no question about its technical feasibility.

Opponents of nuclear energy make several arguments against this form of energy. The first is reactor safety, a problem dramatically demonstrated by the 1986 meltdown of a reactor at Chernobyl, during which errors by the operators caused an explosion and fire in a nuclear reactor. The subsequent release of radioactivity to the atmosphere resulted in an environmental disaster in the region around the reactor as well as radioactive fallout across much of Europe. In addition, nuclear power plants present attractive targets to terrorists, and the prospect of an attack large enough to breach the reactor core and release its radioactive contents to the atmosphere is truly scary.

Another problem is nuclear waste, which is what comes out of the reactor after the nuclear fuel is burned. This waste is extraordinarily radioactive, and it must be safely isolated for many thousands of years. If it were released accidentally, or intentionally in a so-called dirty bomb, the resulting harm in both human cost and ecological damage could be severe. One way to reduce the quantity of waste is to reprocess the fuel, in which usable isotopes of plutonium and uranium are removed and converted back into fuel for another trip through the reactor. Even with reprocessing, though, some waste must be stored for a very long time – and most people do not want the waste to be stored near them.

This leads us to the problem of proliferation. A nuclear bomb requires only a few kilograms of uranium or plutonium. As reactor fuel is mined, enriched, and reprocessed, there exists the possibility that these small amounts of bomb-grade uranium or plutonium could be diverted with the intent of building a nuclear bomb. The diversion could occur by theft from a legitimate nuclear program, or it could be the explicit goal of a rogue state's nuclear program. The net result would be a nuclear weapon in the hands of terrorists or unstable rogue nations, which would present a significant security threat to the rest of the world. This is why the United States and many other countries are so opposed to Iran's development of a nuclear energy program.

Finally, there is the cost. Although nuclear power plants are relatively cheap to run, they are extraordinarily expensive to build. This is one of the primary reasons that no new nuclear power plants have been built in the United States since the 1970s. It may be that nuclear energy cannot be widely deployed without the government playing a key role in financing the construction.

A final option to generate energy without emitting carbon dioxide to the atmosphere is known as *carbon capture and storage,* also known by its initials CCS, or *carbon sequestration.* This refers to a process by which fossil fuels are burned in such a way that the carbon dioxide generated is not vented to the atmosphere. Rather, the carbon dioxide is captured and placed in long-term storage. CCS is a climate-safe technology, but is not renewable (because you are ultimately just burning fossil fuels).

CCS is almost always used in combination with coal combustion, because coal is abundant and produces large amounts of carbon dioxide per joule of energy. An example of a CCS technology is to expose the coal to steam and carefully controlled amounts of air or oxygen under high temperatures and pressures. Under these conditions, atoms in coal break apart and react with the water vapor, producing a mixture of hydrogen, carbon dioxide, and several other gases. The carbon dioxide is separated out, and the other gases are burned in order to generate electricity.

Once captured, the carbon dioxide must be stored. The most likely place to store the carbon dioxide is to inject it deep underground into porous sedimentary rocks, which are widely distributed around the world. Particularly promising sites include depleted oil and gas fields, unminable coal beds, or deep saline formations. This process is technically feasible and would use many of the same technologies that have been developed by the oil and gas industry to enhance the recovery of oil from aging fields. The capacity of these rocks is large enough that they could conceivably hold all of the carbon emitted by human activities.

Using available technology, approximately 85 to 95 percent of the carbon dioxide produced can be captured. This comes at a price, however. A power plant equipped with a CCS system would need to divert 10 to 40 percent of the energy generated into capturing and storing the carbon. Thus, adding CCS would add a few cents per kilowatt-hour of electricity (onto a price in the United States of 10 to 20 cents per kilowatt-hour).

The world has copious reserves of coal, and in our never-ending quest for power, it seems likely that this coal will eventually be burned. CCS may be the only way to simultaneously burn this coal while avoiding climate change. However, although CCS is a promising technology, no large power plant using CCS has ever been built, so the approach remains unproven.

## 11.2.2 Policies to reduce carbon emissions

So the basic question of how to mitigate climate change is really a question of how to encourage the world to switch away from fossil fuels to carbon-free energy sources, such as wind, solar, nuclear, and CCS. Whatever policy we adopt must be able to do this at a sufficiently low cost, with sufficiently little social disruption and without working at cross purposes with other societal goals, such as reducing world poverty and conserving biodiversity, that it can maintain political support over the decades required to address the problem.

Let us begin by addressing the question of why we need regulations to reduce greenhouse-gas emissions in the first place. If people value a stable climate, would not the free market take consumers' interests into account and reduce greenhouse-gas emissions without any government intervention? There is, after all, abundant evidence that the free market is efficient at allocating resources and producing socially beneficial outcomes (although, as the occasional economic meltdown shows, it is not perfect). As U.S. Senator Chuck Hagel said in an interview,[2]

> I have always believed that the marketplace does work. It works because it's based on one fundamental dynamic, which is self-interest of an individual, a company, or a country. The marketplace fosters competition and always trends toward producing a better, cheaper product, which means it is a driver of efficiency. It's in the interests of everyone here to make a cheaper product that's less energy intensive. It cleans up the environment, which has economic advantages too.

So if switching to carbon-free energy is something we want to do, would not the free market take care of this all by itself? The answer is no – emissions reductions sufficient to stabilize the climate will not occur by themselves.

There is, of course, some truth to Hagel's argument. Because consumers have to pay for electricity, consumers have an incentive to reduce how much electricity they consume. This applies pressure to manufacturers to design, build, and sell equipment that uses as little energy as possible. As a result of this market pressure, just about every piece of equipment that you buy today is more efficient than the comparable piece of equipment that was available a few decades ago. That is one of the primary factors behind the world economy's long-term decrease in energy intensity discussed in Section 8.2.

But it is also important to remember that increases in energy efficiency are, by themselves, insufficient to solve the climate change problem. The bulk of emissions reductions will come from a large-scale shift toward carbon-free energy that is not currently occurring. Moreover, there is a good reason it is not now occurring – and may never occur without government intervention.

To understand why, consider the following scenario. Imagine you own a company that produces widgets. Let us assume that it costs your company $1 to manufacture a widget, and during the manufacture process, some carbon dioxide is released into the atmosphere. This carbon dioxide will cause climate change, which causes damages

[2] www.grist.org/article/hagel/.

valued at 10 cents. Thus, the total cost of manufacturing a widget using this process is $1.10. However – and this is important – the widget manufacturer only pays $1; the costs of climate impacts associated with the widget, 10 cents, are borne by everyone in the world. At the same time, the benefit of producing the widget goes entirely to the manufacturer.

Economists call the costs of climate change imposed on the rest of the world by the widget manufacturer an *externality*. More generally, an externality occurs when someone takes an action, and this action imposes involuntary costs on others. To understand the economic implications of an externality, imagine that someone invents a new process for manufacturing widgets, in which it costs $1.05 to produce a widget but no carbon dioxide is emitted to the atmosphere. There are no climate impacts associated with this process, so the total cost is also $1.05 – and this cost is entirely borne by the manufacturer. Because the total cost of building a widget under the new process is 5 cents cheaper than the total cost for the older process, it would be beneficial to society for the widget manufacturer to switch to this new process.

But the widget manufacturer will not switch. The widget manufacturer is only paying $1 per widget under the older method, with the rest of the cost being borne by society. Under the new process, the manufacturer pays the entire cost of $1.05. So even though this new process is cheaper to society as a whole, to the widget manufacturer, this new process is more expensive. Thus, because of the externality, the socially and economically preferred outcome does not occur, a result sometimes referred to as a *market failure*.

Externalities and the associated market failures occur frequently in environmental problems in which some profitable economic activity degrades a common asset such as the atmosphere, the ocean, or a river – and the costs of that degradation are paid for by everyone. In such a situation, the incentive is to utilize the asset without regard to its degradation because those costs are paid by everyone in the society – not the person actually damaging the asset. This is a version of the famous "tragedy of the commons," and it is the fundamental economic explanation for why overfishing is depleting stocks of fish in the oceans, logging is destroying the rainforests, and greenhouse-gas emissions are changing the climate.

In the case of greenhouse gases, emitters exploit the atmosphere by dumping carbon dioxide into it. Because it is free to load the atmosphere with carbon dioxide, the rational behavior of each emitter is to dump as much greenhouse gas into the atmosphere as is necessary to maximize profit. And there is no incentive to reduce emissions. If a company decides to reduce emissions, the entire cost of reducing emissions is borne by the company – while the benefits are spread throughout the society.

Thus, the root economic cause of the climate change problem is that it is free for emitters to dump greenhouse gases into the atmosphere. Therefore, to solve climate change, most economists argue that the costs associated with the emission of greenhouse gases must be shifted back onto the emitter. In the parlance of economics, we need to internalize the externality. This is also frequently described by the principle of "polluter pays." meaning that emitters should be held accountable for the damage they cause.

If the widget manufacturer had to pay 10 cents in order to emit the carbon dioxide associated with the production of a widget under the old manufacturing technology, the manufacturer would be paying the entire cost of manufacturing the widget. In that case, it would be cheaper for the manufacturer to switch to the new method, which costs $1.05 to produce a widget with no emissions of greenhouse gases. This benefits both the manufacturer and society.

**The bottom line on mitigation:** Most experts view mitigation efforts as a necessity. To avoid warming that is generally considered dangerous, emissions have to be reduced below present levels by 50 to 80 percent by the middle of this century.

Although there are many ways to reduce emissions, the only practical way is through technology. Improvements in energy efficiency are important – and make sense even if climate change is not an important problem – but such changes cannot by themselves produce the deep reductions in emissions required. Rather, changes in energy-generation technology are going to be required. This means switching to technologies that generate energy without emitting greenhouse gases, which include solar, wind, biomass, nuclear, and carbon capture and storage.

To encourage the adoption of new technologies, most proposed mitigation policies put a price on emissions. Right now, emitting carbon dioxide to the atmosphere is free, and the costs of the resulting climate change are imposed on everyone in the world. In this situation, there is no incentive for the emitter to reduce emissions of greenhouse gases. Putting a price on emissions makes emitters pay the full cost of their emissions, thereby giving them an incentive to adopt climate-safe energy technology. In Chapter 12, I will discuss pricing carbon in great detail.

## 11.3 Geoengineering

A last solution to the climate change problem is known as geoengineering, which refers to actively manipulating the climate system in order to prevent the climate from changing – or even to return the climate to some past condition. The idea is that we would continue burning fossil fuels and emitting the resulting carbon dioxide into the atmosphere, but we would also make other changes that tend to cool the climate, offsetting the warming from the emissions.

Geoengineering efforts can be roughly divided into two categories: solar radiation management and carbon-cycle engineering. *Solar radiation management* schemes engineer a reduction in the amount of solar energy absorbed by the Earth, $E_{in} = S(1 - \alpha)/4$, where $S$ is the solar constant and $\alpha$ is the albedo (if you do not remember this, you should review Chapter 4).

Most solar radiation management schemes cool the Earth by increasing the albedo. Probably the most frequently discussed way to do this is to inject sulfur dioxide ($SO_2$) into the stratosphere. Once in the stratosphere, this gas reacts with water vapor to form aerosols – liquid droplets that are so small that they have negligible fall speed (discussed in Chapter 6). These aerosols reflect sunlight back to space, thereby increasing the albedo of the Earth and leading to cooling. Injection of sulfur into the stratosphere is the same mechanism by which volcanoes cool the planet.

Another option is to increase the reflectivity of clouds. It turns out that the size of the cloud droplets determines how white a cloud is, with smaller cloud droplets making a cloud whiter or more reflective. This is the same reason that powdered sugar, which is made up of small particles, appears whiter than chemically identical table sugar. Thus, if we could somehow make the particles in clouds smaller, clouds would become more reflective and raise the albedo of the Earth.

One way to do this is to release what are known as cloud condensation nuclei into the cloud. These nuclei serve as seeds that cloud droplets form around. By adding them to clouds, we increase the total number of droplets in the cloud. Because the total water contained in a cloud is basically fixed, this makes the cloud contain more, but smaller, particles. This effect can be seen in ship tracks in the clouds, which were discussed in Chapter 6. As ships steam across the ocean, the exhaust from their diesel engines contains fine particulates that can serve as cloud condensation nuclei. These aerosols are transported by the winds into low-level clouds and brighten them.

The physics supporting these suggestions is robust, and we have high confidence that if any of these schemes were carried out at sufficiently large scale, the planet would indeed cool. There are, however, important disadvantages with these approaches. The first is that solar radiation management schemes focus on temperature, but temperature increases are only one of many impacts associated with climate change – and perhaps not even the most important. We know, for example, that some of the carbon dioxide released into the atmosphere ends up in the ocean, resulting in ocean acidification. Solar radiation management schemes do nothing to address this impact of continued carbon dioxide emissions.

Moreover, solar radiation management schemes may create other problems. The 1991 eruption of Mount Pinatubo, for example, led to substantial changes in global precipitation patterns and an increase in the incidence of drought in some regions. We can therefore expect that reducing the amount of solar radiation reaching the Earth, in addition to cooling the planet, would also lead to changes in the amount and distribution of global precipitation. Whether these changes would be better or worse than climate change is a question whose answer is not clear.

And there are important political problems with this approach. Imagine that a few rich countries in the world (e.g., the United States and Europe) got together to inject sulfur into the stratosphere to cool the planet. Then, China or India experienced a severe drought. Whether the sulfur injection caused that drought or not, the countries affected might well believe that it did. It is easy to imagine that this would lead to a great amount of political tension, possibly even the abandonment of the geo-engineering effort. In the worst case, geoengineering by a group of countries might be considered an act of war by another group of countries that suffer some type of weather-related injury at the same time.

In addition, as long as our society is increasing the amount of greenhouse gas in the atmosphere, geoengineering efforts must be continually strengthened in order to provide an ever-increasing cooling influence to keep the climate stable. So if we were, for example, injecting sulfur into the stratosphere in order to enhance the planet's albedo, then we would have to be injecting ever greater amounts of sulfur over time in order to offset a continually increasing abundance of atmospheric greenhouse gases.

If we ever stopped injecting sulfur into the stratosphere, it would take only a few years for the stratosphere to clear (about the same length of time as it does for the climate to recover after a volcano), during which time the Earth's albedo would be rapidly decreasing. This would cause a significant rise in the Earth's temperature over the following few decades – which would be very bad. Thus, once you start geoengineering, it is difficult to stop.

The second category of geoengineering, *carbon-cycle engineering*, modifies the carbon cycle so that carbon dioxide is more rapidly removed from the atmosphere. As we learned in Chapter 5, it takes a few centuries for the majority of carbon dioxide emissions to be removed from the atmosphere (complete removal takes much longer), and this is the reason carbon dioxide is such a pernicious greenhouse gas. If the lifetime of carbon dioxide could be reduced, it would reduce the amount of carbon dioxide in the atmosphere, thereby reducing the amount of climate change.

Planting trees is an example of carbon-cycle engineering. As the trees grow, they suck carbon dioxide out of the air and sequester it in wood. Another scheme is to add iron to the ocean. Iron is thought to be a limiting nutrient there, so the addition of iron will stimulate the growth of phytoplankton. As the phytoplankton grow, carbon dioxide will be drawn out of the atmosphere and into the ocean. The phytoplankton are then consumed by larger organisms, and subsequent biological activity creates a rain of dead organisms and fecal matter from surface waters into the deep ocean. Thus, adding iron to the ocean has the net effect of drawing carbon dioxide out of the atmosphere and transporting it to the deep ocean.

Another option is to remove carbon dioxide from the air chemically, which is often referred to as *air capture*. This is like CCS, but CCS removes carbon dioxide from the hot exhaust gas of a power plant whereas air capture removes carbon from the free atmosphere. This is an attractive option, but the amount of energy required is staggering, and this severely limits our ability to deploy this technology at scales large enough to remove significant quantities of carbon dioxide from the atmosphere.

In general, carbon-cycle engineering is attractive because, unlike solar radiation management, it does not focus just on temperature. If we balance emissions of carbon dioxide from human activities with removal through carbon-cycle engineering, we will truly stabilize the climate – not just temperature but other aspects, such as ocean pH and precipitation. In fact, a sufficiently aggressive program could lead to a *reduction* in atmospheric carbon dioxide (not just a stabilization), which would eventually undo most of the effects of climate change (but not all – some changes, such as extinction of species, are completely irreversible, whereas others, such as loss of the world's largest ice sheets, are effectively irreversible on any time scale that we care about).

There are, of course, some problems with the various approaches to carbon-cycle management. For example, real engineering of the carbon cycle, such as adding iron to the ocean, is risky. Because of significant uncertainties in our knowledge of how carbon cycles in the ocean, there is a possibility that that these schemes will not work. In other words, we might add iron to the ocean only to find out that, because of unanticipated physics or biology, no extra carbon dioxide was removed from the atmosphere. Even worse, it might have unforeseen and serious impacts on ocean ecosystems. Thus, in trying to address climate change, we may cause an entirely new environmental problem – and not even solve the problem we were intending to address.

Overall, geoengineering has several general qualities that make it an attractive policy response to climate change. First, and possibly most importantly, it attacks the climate change problem through technology but does not seek to limit emissions in any way. That allows us to continue doing exactly what we are doing now: burning fossil fuels and consuming energy as fast as we possibly can. Second, geoengineering is a relatively rapid response. Many schemes can be implemented in a decade or so. This contrasts favorably with mitigation, for which emissions reductions must begin now in order to head off climate changes that will occur in 50 to 100 years. Thus, like adaptation, geoengineering is often thought of as something that we will do in the future, if climate change turns out to be bad.

Geoengineering also avoids many of the moral problems of adaptation. A successful geoengineering effort will help the entire planet, not just the rich countries. Moreover, the cost estimates are often cheap enough that one or a few rich countries could shoulder the burden. However, it also raises moral problems of its own: Should we be engineering the climate? Who decides what the ideal temperature is? What happens if geoengineering efforts harm one particular region while benefiting all others? These are all difficult questions that have to be resolved before the world uses geoengineering to address climate change.

**The bottom line on geoengineering:** Geoengineering is an appealing but risky approach to dealing with climate change. Although it may work, the risk exists that geoengineering may lead to unintended consequences that would leave the world worse off. And some approaches do not address all of the impacts of climate change and may only be viable for a relatively short period of time (less than a century). Thus, in a world where climate change is handled responsibly through mitigation and adaptation, it is unlikely that geoengineering would be needed.

Nevertheless, it is easy to imagine a future in which the world makes no progress in reducing emissions. If, by the middle of the twenty-first century, emissions are large and climate change is out of control, geoengineering may represent the last hope for avoiding truly disastrous climate change. It is in this type of scenario that the deployment of geoengineering is a reasonable strategy. But even here, geoengineering is unlikely to be the final solution. Rather, it will be a way to buy time to let emergency mitigation efforts reduce emissions.

## 11.4 Chapter summary

- Responses to climate change can be roughly divided into three categories: adaptation, mitigation, and geoengineering.
- Adaptation means learning to live with climate change; mitigation refers to reducing emissions of greenhouse gases, thereby preventing the climate from changing in the first place; and geoengineering refers to active manipulation of the climate system in order to engineer a cooler climate.

- Because not all climate change can be prevented, we will have to adapt to some climate change. However, there are both moral and practical problems with relying on adaptation as our only response.
- Mitigation refers to efforts to reduce emissions, thereby preventing future climate change. Most experts and world leaders view mitigation as a vital component of any plan to address climate change. This will be accomplished mainly by transitioning from fossil fuels to energy sources that do not emit greenhouse gases. A key part of policies designed to encourage transition to climate-safe energy is to put a price on emissions of greenhouse gases, which are presently free.
- Geoengineering is an active manipulation of the climate system. Under this approach, our society would continue adding greenhouse gases to the atmosphere, but we would intentionally change some other aspect of the climate in order to cancel the warming effects of the greenhouse gases. Geoengineering strategies can be broken into two categories – solar radiation management and carbon cycle engineering. Because of potential problems with geoengineering, it is generally considered a last-ditch approach, only to be used if adaptation and mitigation approaches fail.

# Additional reading

Much has been written about our policy response to climate change in the form of books, reports, and scholarly publications. Many are available online, and you can find them by searching the Internet. Here are a few of particular note.

The IPCC's Working Group II covers impacts and adaptation, and it remains one of the most authoritative summaries of what we know. You can download and read the most recent report at ipcc-wg2.gov/AR5/report/. The IPCC's Working Group III focuses on mitigation of climate change. You can download the most recent report from this working group at mitigation2014.org/.

For a more U.S.-centric view of adaptation and mitigation, see these reports: National Research Council, *Adapting to the Impacts of Climate Change* (Washington, DC: The National Academies Press, 2010); available online at www.nap.edu/catalog .php?record_id=12783 and National Research Council, *Limiting the Magnitude of Future Climate Change* (Washington, DC: The National Academies Press, 2010); www.nap.edu/openbook.php?record_id=12785.

The key to understanding why we have a climate problem is recognizing that there are "hidden" costs to energy that are not paid by the consumers, known as externalities. This comprehensive report details these surprisingly hefty costs (National Research Council, *Hidden Costs of Energy: Unpriced Consequences of Energy Production and Use* [Washington, DC: The National Academies Press, 2010]; available online at www.nap.edu/catalog.php?record_id=12794).

D. J. C. MacKay, *Sustainable Energy – Without the Hot Air* (Cambridge: UIT Cambridge, 2009). This is a sober and quantitative analysis of how hard it will be to replace fossil fuel energy with carbon-safe energy sources. The short answer is that it would not be easy (download the book at www.withouthotair.com).

There are several good and very readable books about geoengineering out there. These include E. Kintisch, *Hack the Planet: Science's Best Hope – or Worst Nightmare-for Averting Climate Catastrophe* (New York: Wiley, 2010) and J. R. Fleming, *Fixing the Sky: The Checkered History of Weather and Climate Control* (New York: Columbia University Press, 2012).

See www.andrewdessler.com/chapter11 for additional resources for this chapter.

# Terms

Adaptation
Air capture
Carbon capture and storage
Carbon-cycle engineering
Carbon-free-climate-safe energy sources
Carbon sequestration
Externality
Geoengineering
Market failure
Mitigation
Renewable energy
Solar photovoltaic
Solar radiation management
Solar thermal

# Problems

1. Our responses to climate change can be put into three general categories. List the categories. For each category, give one example of an action that would fall into that category.
2. a) What are carbon-free energy sources? List the ones discussed in the book.
   b) Is carbon-free energy the same as renewable energy?
   c) Is nuclear energy carbon free? Is it renewable?
3. Why do economists generally believe that the free market will not solve the climate problem by itself?
4. What is an externality?
5. Do some research and find an example of "the tragedy of the commons." Explain how global warming is an example of this type of problem?
6. a) Your friend says that "we should rely entirely on adaptation as our response to climate change." Is this a good idea?
   b) I argued here that adaptation must be at least part of our response. Why?
7. a) Explain one way we can "geoengineer" a higher planetary albedo.
   b) Explain one way we can "geoengineer" a reduction in carbon dioxide.

8. a) In this chapter, we divided geoengineering approaches into two categories. What are they?

   b) What are the advantages and disadvantages of geoengineering?

   c) I argued that that geoengineering should be used in what circumstances?

9. a) Imagine a credit card whose bill was divided up and sent to everyone in the United States (i.e., if you purchased something on this card, every person in the United States would get a bill for 1/300,000,000th of your total cost). Would the average person spend freely with this credit card? Or would they be as thrifty as they would if they had to pay the entire bill?

   b) Now imagine that every person in the United States has a credit card like this. What do you think is going to happen?

   c) How is this situation related to the climate change problem?

10. In this chapter, we explored the terms of the IPAT relation and concluded that reducing emissions could really be achieved only through reduction of one term. Which term is it, and why is that our only real option?

11. The technology term in the IPAT relation can be further divided into two terms.

    a) For each term, give an example of a technological switch that reduces it.

    b) One of these terms is the key to deep reductions in emissions. Which one is it? What kinds of changes are required to make such deep reductions?

12. Why do mitigation policies have little ability to influence the climate over the first half of the twenty-first century?

13. In this chapter, we talked about negative externalities. Can you think of an example of a positive externaility?

# 12

# Mitigation policies

Chapter 11 discussed the three options we have to address climate change: adaptation, mitigation, and geoengineering. Adaptation will, by necessity, be an important part of our response to climate change. However, relying entirely on adaptation as our only response to climate change is fraught with problems. Geoengineering is another possibility, but one that few people think should be used now. Rather, it is the last resort – like an airbag in a car – that you turn to if other approaches to address climate change fail.

The remaining option is mitigation – the reduction of greenhouse gases emissions so as to avoid climate change – and there is agreement among those who have looked seriously at the problem that we should embark on mitigation efforts right now. Mitigation schemes will have little effect on the climate of the next few decades, but a successful mitigation effort would allow us to avoid large climate changes in the second half of this century and beyond.

As we learned in Chapter 11, reducing greenhouse-gas emissions requires improvements in the energy efficiency of our economy and, most importantly, converting our energy system to one that primarily utilizes carbon-free energy sources, such as solar, wind, nuclear, and CCS. We also discussed in the last chapter why these emissions reductions would not occur without explicit government policy. In this chapter, we explore in detail the policy options that governments can use to reduce emissions.

## 12.1 Conventional regulations

The conventional approach to regulation, often described colloquially as *command-and-control* regulation, requires all emitters in a particular economic sector to meet a single standard. Electricity companies, for example, might be required to generate energy by using a particular technology, such as wind or CCS. Cars might be required to be gas/electric hybrids. Alternatively, regulations may limit total emissions of a pollutant, or enforce a standard of greenhouse gases emitted per kilowatt-hour generated (for power plants) or greenhouse gases emitted per mile driven (for cars). For example, in 2012 the Obama Administration proposed requiring new power plants to emit less than 1,000 pounds of carbon dioxide per megawatt-hour of energy produced. Burning coal produces more than this, so this regulation effectively prohibits the construction of new coal-fired power plants (without CCS).[1]

---

[1] This proposal was withdrawn in late 2013 when the Obama Administration released a more comprehensive proposal for regulations that would also cover existing power plants.

The conventional approach has the advantage that it is clear and easy to understand. Even today, many environmental regulations, including regulations on air pollution, fall into this category. Since the 1980s, however, weaknesses with this approach have been identified, and it has been falling out of favor with regulators. First, technologies specified (e.g., wind, CCS) may not actually turn out to be the best ones. Second, the regulations force all emitters to meet the same emissions standards. This ignores the fact that some emitters can reduce emissions more cheaply than others. Third, conventional regulations provide no incentive for the development and deployment of new technologies that reduce emissions beyond the specified target.

Because of these disadvantages, there is little talk in policy circles of attempting to solve the climate change problem primarily with conventional regulations. Instead, market-based regulations are now preferred, and I will spend the rest of the chapter discussing them.

## 12.2 Market-based regulations

In Chapter 11, I discussed why the free market is unlikely to solve climate change without intervention from the government: The basic reason is that it is free to dump greenhouse gases into the atmosphere. The cost of these emissions is imposed on everyone, not the emitter — which is why economists call this an externality. Because the costs of the emissions are not borne by the emitter, there is no economic incentive for the emitter to reduce emissions. The solution therefore is to make emitters pay for emitting greenhouse gases. In so doing, we give the emitters an economic incentive to reduce emissions.

Making emitters pay for their emissions is a market-based solution. It does not tell anyone how much they can emit or what technology to use – it only requires them to pay for whatever emissions they do make. And it makes reductions more cheaply than convectional regulations. In the next two sections, I will discuss the two market-based regulatory approaches most frequently discussed in the climate change policy debate: carbon taxes and cap-and-trade systems.

### 12.2.1 Carbon tax

The first approach is a *carbon tax*. Under this policy, emitters have complete freedom to emit as many tons of greenhouse gas to the atmosphere as they choose, as long as they pay a specified fee to the government for each ton released to the atmosphere.

To understand why a carbon tax reduces emissions, let us imagine a power plant, which we will call Plant A, that emits 10 tons of carbon dioxide into the atmosphere each year. The third column of Table 12.1 contains the *marginal cost* for reducing emissions for Plant A, which is the cost of reducing a particular ton of emissions. Thus, Plant A can reduce its annual emissions by 1 ton – from 10 tons to 9 tons – for $1. Reducing emissions another ton, from 9 tons to 8 tons, costs an *additional* $2. Thus, the total cost of reducing emissions from 10 to 8 tons is the sum of the marginal

| Emissions reduced by (tons) | Units emitted (tons) | Plant A's cost ($) | | Plant B's cost ($) | |
|---|---|---|---|---|---|
| | | Marginal | Total | Marginal | Total |
| 0 | 10 | – | – | – | – |
| 1 | 9 | 1 | 1 | 2 | 2 |
| 2 | 8 | 2 | 3 | 4 | 6 |
| 3 | 7 | 3 | 6 | 6 | 12 |
| 4 | 6 | 4 | 10 | 8 | 20 |
| 5 | 5 | 5 | 15 | 10 | 30 |
| 6 | 4 | 6 | 21 | 12 | 42 |
| 7 | 3 | 7 | 28 | 14 | 56 |

**Table 12.1** Cost of reducing emissions for Plants A and B

costs,[2] that is, $1 + $2, for a total cost of $3. The next ton of emissions costs $3 to eliminate, so the total cost of reducing emissions from 10 to 7 tons is $1 + $2 + $3 = $6. And so on.

The marginal cost of reducing emissions increases as emissions are progressively reduced. To see why, think about golf. When you just start out, you may be shooting 130 strokes over eighteen holes. It takes relatively little effort to reduce your score by one stroke, down to 129 – perhaps only 1 hour at the driving range. Taking another stroke off your score, reducing your score to 128, takes slightly more work, maybe two more hours at the driving range. By the time your score reaches 80 and you are getting close to par, it can be extremely difficult and take enormous practice to reduce your score by one stroke to 79. Eventually, as you approach and possibly surpass par, you reach a limit beyond which you will never move past, no matter how hard you work.

In this golf example, the marginal cost is the amount of effort you have to apply to reduce your golf score by one stroke. The fact that the amount of work it takes to improve your score by one stroke increases as you get better is a version of the law of diminishing returns: each additional unit of effort (e.g., an hour at the driving range) produces less of an improvement than the previous unit of effort.

For Plant A, fine-tuning the machinery in the plant may reduce the first ton at very little cost. Once the equipment has been tuned up, however, additional reductions are harder to make. Eliminating the second ton may require replacing some outdated equipment with newer, more efficient equipment. This will cost more than the first ton. The third ton may require even more equipment replacement, or perhaps wholesale changes in the plant's operation. This ton will therefore be more expensive to eliminate than the previous two tons.

Now imagine that a carbon tax of $4 per ton is imposed on the emitters – meaning that for every ton that is emitted into the atmosphere, the plants have to pay the government $4. How would each plant respond? Remember that Plant A has total freedom to emit as much as it wants – the carbon tax does not specify any reduction. Plant A will therefore search for its cheapest alternative.

[2] For those of you who know calculus, you can think of the marginal cost as the derivative of the total cost function and the total cost as the integral of the marginal cost function.

Plant A can emit the tenth ton and pay a tax of $4, or it can pay $1 and not emit that ton. It does not take a financial genius to conclude that the rational thing to do is to not emit the ton. Now emissions are down to 9 tons, and Plant A can emit the ninth ton and pay a tax of $4, or it can pay $2 and not emit that ton. For this ton, too, the rational thing to do is to not emit that ton. Now emissions are down to 8 tons, and Plant A can emit the eighth ton and pay a tax of $4, or it can pay $3 and not emit that ton. Again, the rational thing to do is to not emit that ton.

Now things get a bit trickier. For the 7th ton, Plant A can emit the ton and pay a tax of $4 or it can pay $4 and not emit that ton. From a purely financial point of view, these two alternatives are equivalent. I suspect, though, that most companies will reduce that last unit, because it is the same cost as the tax and the company can use this to burnish its environmental reputation. So we can assume that Plant A will choose to not emit the seventh ton.

For the sixth ton, Plant A can emit the ton and pay a tax of $4 or it can pay $5 and not emit that ton. The rational thing to do in this situation is to pay the tax and emit that ton. Thus, under a carbon tax of $4, Plant A will reduce emissions by 4 tons.

Now let us consider a second plant, which we will call Plant B, whose marginal costs are also listed in Table 12.1. Plant B can emit the tenth ton and pay a tax of $4, or it can pay $2 and not emit that ton. Clearly, Plant B will not emit the ton. Now emissions are down to 9 tons, and Plant B can emit the ninth ton and pay a tax of $4, or it can pay $4 and not emit that ton. As for Plant A, we can assume that Plant B will choose to not emit that ton. Now emissions are down to 8 tons, and Plant B can emit the eighth ton and pay a tax of $4, or it can pay $6 and not emit that ton. Here, the rational thing to do is to pay the tax and emit that ton. Thus, under a carbon tax of $4, Plant B will reduce emissions by 2 tons.

Thus, Plant B reduces emissions less than Plant A does for the same carbon tax rate. The reason is that Plant B has higher marginal costs for reducing emissions. This may arise for any number of reasons – for example, Plant B may be older than Plant A and so be using outdated technology that is not amenable to reducing emissions.

*To summarize, under a carbon tax each emitter will reduce emissions until the marginal cost of reduction is equal to the carbon tax rate.* For Plant A, the marginal cost equals the tax rate of $4 per ton when emissions have been reduced 4 tons, whereas for Plant B, the marginal cost equals the tax rate when emissions have been reduced by 2 tons. Because marginal costs vary among emitters, some emitters will make deeper cuts than others.

The total reduction in emissions from the two plants in response to a carbon tax of $4 per ton is 6 tons: a reduction of 4 tons from Plant A and 2 tons from Plant B. The total cost to Plant A of reducing emissions is $1 + $2 + $3 + $4 = $10, whereas the total cost to Plant B is $2 + $4 = $6. Thus, the total cost to society of reducing 6 tons of emissions is $10 + $6 = $16.

Under a conventional command-and-control approach, there is a single performance target that each plant is required to meet. So an emissions reduction of 6 tons might be achieved, for example, by having both Plant A and B reduce emissions by three tons. This would cost Plant A $1 + $2 + $3 = $6 to reduce 3 tons, whereas it would cost Plant B $2 + $4 + $6 = $12 to reduce 3 tons. The total cost of this 6-ton reduction is $18.

This is an important result. The carbon tax of $4 per ton resulted in a 6-ton reduction for $16, whereas the conventional command-and-control approach resulted in a 6-ton reduction for a cost of $18. The carbon tax is cheaper because of its *flexibility* – it shifts reductions to the lowest marginal cost emitters, in this case, Plant A – so that the emissions reductions are made where they are cheapest, which lowers overall cost to society.

A carbon tax would be reasonably easy to implement. Most greenhouse gases come from fossil fuels, and these are produced at a relatively small number of sites. A carbon tax could be applied to the fossil fuel when it is extracted from the ground, using the administrative infrastructure for existing taxes, such as excise taxes on coal and petroleum. The price of the tax would then follow the fuel through the market, where the end user would finally pay it. A tax credit would be generated if the carbon is used in such a way that it was not released into the atmosphere (such as production of plastic or capture of carbon in coal combustion followed by sequestration).

As part of a long-term policy, the carbon tax would start out relatively small and, over several decades, gradually increase until emissions have been reduced to the target level. Gases other than carbon dioxide, such as methane or nitrous oxide, would also be taxed but at a rate that takes into account how effective each one is at warming the planet. For example, 1 ton of methane contributes approximately twenty times more warming than a ton of carbon dioxide, so the tax on methane should be proportionately higher than the tax on carbon dioxide.

The costs of reducing emissions would eventually be passed on to consumers. Thus, the net effect of a carbon tax is to raise the prices of goods and services by an amount proportional to the amount of the greenhouse gases released. Goods and services that are produced with little or no emission of greenhouse gases will not experience price increases, whereas the costs of goods and services that require the emission of significant amounts of greenhouse gases may see large price increases.

Many people automatically consider taxes to be bad, so they look at suggestions of a carbon tax with, to put it mildly, disdain. However, most economists argue that a well-designed tax serves a useful economic purpose. For activities that generate negative externalities (costs imposed on society, such as emitting greenhouse gases or smoking cigarettes), the free market prices these activities too low, leading to overconsumption of the associated good or service. Taxes on these activities correct for this and reduce consumption, which produces a socially beneficial outcome. Thus, an economist thinks of a carbon tax as fixing a problem in the free market. Unfortunately, given the automatic opposition that new taxes generate, even those that make economic sense, the present prospects for implementing a carbon tax in the United States and in many other countries are dim.

## 12.2.2 Cap and trade

An alternative way to put a price on greenhouse-gas emissions is a *cap-and-trade* system. Under cap and trade, the government issues a fixed number of permits each year, with each permit allowing the holder to emit a fixed amount (often 1 ton) of greenhouse gas to the atmosphere. Emitters must hold permits for the amount of greenhouse gas they emit to the atmosphere. Thus, the total number of permits issued sets a cap on total emissions. Emitters with extra permits can sell them to those needing additional permits (hence the *trade* part). The price of the permits is set by the market, not by the government.

The economics of a cap-and-trade system is nearly identical to the economics of the carbon tax. If the marginal cost of reducing 1 ton of greenhouse gas emissions is less than the cost of the permit, the emitters will not emit that ton. This allows them to either avoid having to buy a permit, or, if they already have a permit for this ton, they can sell it at a profit. If the marginal cost is more than the permit, the emitters will acquire a permit and emit that ton. In the end, the emitters will reduce emissions until the marginal cost of reducing emissions equals the price of the permits.

So if permits cost $4 per ton, Plant A will use permits to emit 6 tons, thereby reducing emissions by 4 tons. Plant B will use permits to emit 8 tons, thereby reducing emissions by 2 tons. This is the same result that was obtained for a carbon tax of $4 per ton.

Under most cap-and-trade systems, when a unit of greenhouse gas is emitted, a permit is retired. Therefore, the government must continually issue new permits to replace those that have been used. Over several decades, the number of permits issued each year will decrease following a prescribed schedule until the target emissions level is reached.

One of the most contentious issues in any cap-and-trade system is how the government issues those permits. One approach is for the government to auction the permits off. In that case, companies would buy the permits from the government and then pass the cost of the permits on to their customers through higher prices for their products. This approach has the advantage that permits go to those emitters who value the permits the most – and are therefore willing to pay the most. These will be the highest marginal cost emitters, for whom emissions reductions are most expensive.

Emitters, like utilities that burn coal, oppose auctioning the permits because they would have to pay the cost of the permits, which would then be passed on to consumers in the form of higher prices. This will reduce demand for their product – which would cost them money. The alternative is for the government to give away permits to companies for free. Because permits can be sold, this is equivalent to the government giving emitters money. Unsurprisingly, emitters favor this approach.

When giving the permits away for free, the decision about how to allocate permits is not determined by the market, like they would in an auction, but by other issues, such as fairness and political connections. For example, the imposition of a cap-and-trade system will be potentially disruptive to industries that emit a lot of carbon to the atmosphere (e.g., coal companies). Giving these industries free permits essentially provides financial aid to help them adjust to a new world in which emitting carbon to the atmosphere is no longer free. Politicians can also distribute permits to curry favor from particular constituents or to buy support for the policy from particular industries.

In the most recent bill considered by the U.S. Congress in 2009 and 2010, for example, the majority of permits would initially be given away for free. Many of these free permits would go to many large emitters, including coal companies. Because of that, many of these large emitters strongly favored passage of the bill. The bill then required a slow transition to auctioning 100 percent of the permits over the next two decades.

One last issue is that both a carbon tax and a cap-and-trade system (with auctioned permits) would create an enormous transfer of wealth from consumers to the government. Some of this wealth could be used to help those with low incomes, who would be disproportionately hurt by the rise in energy prices. The government could also use the income for other beneficial activities, such as research and development of new energy technology or reduction of the deficit. One frequently made suggestion is to use this money to reduce other taxes, such as those on labor and capital, or rebate the money evenly to every citizen (sometimes called "cap and dividend").

## 12.2.3 Carbon tax versus cap and trade

Carbon tax and cap-and-trade systems are quite similar in many ways. Both reduce emissions by putting a price on emissions. Both systems allow companies to emit as much as they want, as long as they pay the tax or possess a permit for each unit emitted. In doing so, both move emissions reductions to where they are cheapest, namely to the lowest marginal cost emitters. In both cases, the emitter reduces emissions until the marginal cost of reducing the next ton of emissions is equal to the price on emissions.

Putting a price on emissions means that both approaches raise the price of fossil fuels and the goods and services made from them in proportion to the amount of greenhouse gases emitted by their consumption. Although consumers may not like to see prices go up, economics tells us it is the most efficient way to reach a socially optimum level of emissions, and it does so through several mechanisms. First, higher prices encourage consumers to reduce their consumption of greenhouse-gas-intensive goods and services. Second, putting a price on emissions encourages the economy to substitute climate-safe technology for their present technology. This occurs because these policies raise the prices of emitting greenhouse gases, so they increase the economic competitiveness of climate-safe technologies (e.g., solar, wind, nuclear). Third, and most importantly, it encourages research on and development of new technologies that can replace today's technologies that produce greenhouse gases. Humans are amazingly clever, and putting a price on emissions signals to the market that innovation and breakthrough technologies will pay off handsomely. With this market incentive, we can expect new technologies that could dramatically reduce the cost of stabilizing the climate.

However, there are some important differences between these two approaches. Under a carbon tax, the policymakers set the tax rate, which in turn sets the cost to society of the emissions reductions. But it is not exactly known what the economy's marginal cost of reduction is, so this means there is uncertainty in exactly how much of an emissions reduction will occur given a particular tax rate. Under a cap-and-trade system, in contrast, the policymakers set the total number of permits issued, and therefore the total emissions from the economy. However, the uncertainty in the marginal costs means that it is unknown how much it will cost to reach the specified level of emissions.

Here is a simple analogy. Imagine you go to a store to buy some soda. You are given $50 and instructions to buy as much soda as you can. In that case, you know the total cost you will incur upfront ($50), but you do not know how much soda this will purchase. This is analogous to a carbon tax. This creates the potential problem that the tax rate set by the government will not achieve the desired emissions reductions. However, the importance of this problem is frequently overstated. Because emissions reductions will take several decades to reach the desired level, the tax rate can be adjusted over time so that the desired emissions levels are eventually reached.

Now imagine that you go to the store and are given instructions to buy twelve cases of soda. In this case, you know exactly how much soda you will get (twelve cases), but you do not know the cost. In the worst-case scenario, you might significantly underestimate the cost of a case of soda and not take enough money – and thereby be unable to get all twelve cases. This is the situation with a cap and trade. The total number of permits issued by the government sets the limit on emissions. However, the cost is uncertain. This creates a potential problem for cap-and-trade systems: Policymakers will issue too few permits (in an effort to bring emissions down sharply), and the cost of complying will be so high that significant economic disruption occurs. If this happens, the program might lose political support and be abandoned. Because of this risk, many cap-and-trade systems have tended to err on the side of caution and issue far too many permits. In such a case, the price of the permits goes to zero and little or no emissions reductions occur.

To address the problem of runaway costs in a cap-and-trade system, a better approach is to implement an *escape valve*. Should the cost of permits rise above a predetermined threshold, the government will sell more permits at that predetermined price. This would loosen the cap and increase emissions, but it would also reduce the cost to the economy. If, in contrast, the government issues too many permits (in an effort to keep costs down) and the price of permits drops below a predetermined floor, the government will commit to buying all permits being sold at that price, thus preventing the price of the permits from falling below that floor.

For political reasons, cap and trade has generally been the preferred climate policy for the last two decades. The European Union has a cap-and-trade system operating today, which will be discussed later in this chapter. In the United States, however, political and economic events during the first term of the Obama presidency have made cap and trade a toxic commodity, just like a carbon tax, and there seems to be no chance that an economy-wide one will be implemented at the federal level. In fact, at present it appears unlikely that the United States Congress will implement any comprehensive emissions-reduction policy – in the next few years, at least.

## 12.2.4 Offsets

Imagine that you own a power plant that emits 100 tons of carbon dioxide into the atmosphere every year. Imagine that you also own a forest, which absorbs 100 tons of carbon dioxide from the atmosphere each year. If a carbon tax is implemented, how much tax should you pay? Do you pay a tax on emissions (100 tons, from the power plant) or do you get credit for the carbon dioxide removed from the atmosphere by the forest? If you get full credit for the forest, then your net emissions are zero because the emissions from the plant are canceled by the uptake by the forest and you owe no carbon tax.

Actions that reduce the amount of carbon dioxide in the atmosphere – you can think of these as "negative emissions" – are often referred to as *offsets*. They are called offsets because, if credit is given to these negative emissions in a climate policy, emitters can use them to offset their greenhouse-gas emissions, thereby reducing their carbon tax.

From a physics standpoint, offsets *should* count as negative emissions. There is no difference to the climate system between not emitting 1 ton of carbon dioxide and emitting 1 ton while, at the same time, removing 1 ton via an offsetting mechanism. Offsets make sense from an efficiency standpoint, too. In much the same way that carbon tax and cap-and-trade systems are efficient because they encourage the lowest marginal cost emitters to make the reductions, offsets provide further flexibility in exactly how emissions are reduced. It may be cheaper, for example, for a coal-fired power plant to offset emissions by planting trees than it would be for it to capture the carbon produced in the coal combustion. Because of this, offsets would be expected to lower the total cost of reaching any specified emissions target.

However, offsets are a much dicier proposition in reality. First, many offsets are difficult to verify. For example, measuring carbon uptake by a forest is an extremely complex problem. And what happens if the forest grows for several years, and then a forest fire burns it to the ground, releasing the sequestered carbon dioxide back to the atmosphere? How is this accounted for? Does the forest owner have to refund the payments he received for the offsets?

Then comes the question of *additionality*: would the offsetting action have taken place without the additional value given to the offsetting action by the carbon emissions regime? To understand what I mean by this, consider the following example. You own a plot of land, so you go to a local power plant and offer to plant trees on it if they pay you. They do so, and in turn they use the carbon absorbed by the growing trees to offset some of their emissions, which reduces their carbon tax.

The problem arises because we do not know what would have happened without the payment from the power plant. In order for these offsets to actually reduce carbon in the atmosphere, the offsetting action (in this case, planting of trees) should only have occurred because of the payment from the power plant. If you would have planted those trees anyway, then the payments from the power plant did not lead to any reduction in carbon in the atmosphere and should not be used as offsets.

In other words, for an offset to count, we must be sure that the offsetting actions are *in addition* to what would have happened anyway and would not have occurred without the value that they have for climate change avoidance. Otherwise the offsets achieve no environmental good. Mitigation programs that include offsets must therefore establish a mechanism to determine whether an offset satisfies additionality.

For these reasons, offsets turn out to be one of the most controversial aspects of any mitigation program. In fact, some of the biggest stumbling blocks in the negotiation of the Kyoto Protocol were the proposals to allow offsets from forests and agricultural lands to satisfy a major part (between one-fourth and one-half) of the total emissions reductions of each country. This proposal was pushed by the United States (a country with a lot of forest and farm lands), but it was steadfastly rejected by Denmark and Germany.

An aside: The European Trading System

In response to their Kyoto Protocol obligations, the European Union (EU), an economic and political union of more than two-dozen European countries, created the EU Emissions Trading System (ETS) in 2005. The basic structure of the ETS is a cap-and-trade system, and it covers factories, power plants, and other major emitters, which make up 40 to 50 percent of the EU's emissions.

The ETS has gone through several phases since its inception. During the first phase (2005–2007), almost all permits were given to businesses free of charge. Because of a lack of knowledge of historical emissions and worry that an aggressive cap on emissions might hurt the economy, the number of permits issued exceeded actual emissions by a large margin, leading to a price for the permits of zero. Given that emitters reduce emissions until the marginal cost of the next unit of reduction equals the permit price, a permit price of zero yielded no incentive for companies to reduce emissions and effectively no emissions reductions.

The second phase of the ETS covered the Kyoto Protocol's commitment period, 2008 to 2012. During this phase, most permits were still given away free. But in response to the low permit price during the first phase, the number of permits issued was scaled back. However, 2008 also coincided with the beginning of a worldwide economic recession, which greatly reduced energy consumption – and therefore emissions. As a result, even the reduced number of permits was too high, and the permit price declined over this period from about $30 per ton to about $6 per ton – again, the low permit price provided little incentive to reduce emissions. Emitters were also allowed to use offsets and related devices to purchase emissions reductions in other countries and use those in place of reducing their own emissions.

The third phase of the ETS covers 2013 to 2020. Permits will primarily be issued via auctions, which will help distribute the permits to those who value them most. There were also a number of other changes designed to increase the efficiency of the emissions reductions. Despite these changes, the price of permits has remained stubbornly low.

Initial problems like those of the ETS are inevitable in any program as complex as a carbon trading system. And, over time, they are being worked out. Other cap-and-trade systems are using the experience of the ETS to help them avoid its mistakes. For example, California set up a cap-and-trade system in 2013 as part of a state-level effort to reduce emissions. In order to avoid issuing too many permits, California regulators studied emissions levels for several years to determine what the correct number of permits would be. Permits are issued by a mix of free allocation to certain industries judged in need of help during the transition to a low-carbon economy and auction to everyone else. To avoid the price of permits falling too low, California established a minimum price of around $11 per ton for the auctioned permits. During the auction, the price of the permits ended up slightly above this minimum price.

Over time, as more of these trading systems are implemented, we can expect the world to gain important experience on the best practices of emissions trading systems.

# 12.3 Information and voluntary methods

A final way to reduce emissions is simply to give people information. If people can be convinced that climate change is a serious problem, and then provided ways to address the problem, they may take some action to address it without any further prompting by the government.

Information can indeed affect purchasing decisions. For example, car dealerships in the United States are required by law to put mileage stickers on cars they are selling. Although not every car buyer is concerned with mileage, many are, and this information helps them make the socially beneficial decision of buying a high-mileage car.

In the case of climate change, an example of relevant information is a greenhouse-gas registry. The requirement to simply report emissions can provide strong incentives for companies to reduce their emissions. Companies whose emissions far exceed those of their competitors will be embarrassed, whereas those with low emission may be viewed as socially responsible and thereby favored in the marketplace. In both cases, a registry will give companies incentive to reduce their emissions.

However, this approach has limits. Although informational and voluntary approaches are quite useful and can be effective at encouraging people to make some changes to their behavior, these approaches generally do not compel people to make large or difficult changes. For example, imagine that your professor asks you to work a few extra problems (that will not be graded) before the exam. Given the expectation that working the problems may help you on the exam, and that it is not a large investment in time, you might choose to do so. Now imagine that your professor asks you to write a twenty-five-page term paper. If the paper represents a large fraction of your grade, most students will take the assignment seriously. However, if your professor informs you that she is not going to grade the term paper, many students will not put much effort into the paper.

So the government can provide information about climate change and how to reduce emissions, and it may well cause some people to make some changes. However, the large changes necessary for us to stabilize the climate are too big to be motivated simply because we have been told we *ought* to make those changes. Thus, informational and voluntary approaches will likely form part of our response to climate change. They will not, however, form the fundamental basis for our approach to reduce emissions.

# 12.4 Chapter summary

- The central pillar of most mitigation policies is putting a price on emissions of greenhouse gases. There are two primary policies to do this: a carbon tax and a cap-and-trade system.
- Under a carbon tax, emitters must pay a tax for each unit of greenhouse gas emitted. Under a cap-and-trade plan, each emitter must hold government-issued

permits equal to the amount of greenhouse gases emitted; extra permits can be traded.

- Over decades, the tax rate will rise or the number of permits issued will decrease following a predetermined schedule until the desired emissions level is reached.

- Under these policies, emitters reduce emissions until the marginal cost (the cost of reducing the next unit) is equal to the carbon tax or the price of the permit.

- These policies are efficient because they shift emissions reductions to where those reductions can be made most cheaply.

- Offsets are processes that remove carbon from the atmosphere – they can be thought of as negative emissions. Whether these are allowed to offset real emissions is one of the most contentious parts of emissions-reduction policy debates. Offsets should satisfy additionality for them to count. This means that the offsetting activity would not have occurred without the additional value of the activity from its impact on emissions.

- Because of the long lifetime of carbon dioxide, as well as the time it takes for mitigation policies to reduce emissions, mitigation efforts we begin today will significantly affect the climate only in the second half of the twenty-first century.

# Terms

Additionality
Cap and trade
Carbon tax
Command-and-control regulation
Escape valve
Flexibility
Marginal cost
Offsets

# Additional reading

There has been a huge amount written about both carbon taxes and cap and trade. A quick Google search will turn up more books, magazine articles, and whitepapers than you could ever read. So dive in! What follows here are a few particular suggestions.

P. Krugman, "Building a Green Economy," *New York Times Magazine*, April 7, 2010. This is a clear and concise summary of the economics of climate change policy. It pulls together many of the concepts from Chapters 11 and 12 (download at www.nytimes.com/2010/04/11/magazine/11Economy-t.html).

W. D. Nordhaus, *The Climate Casino* (New Haven, CT: Yale University Press, 2013). This is a much longer, but also more wide-ranging, discussion of the economics of climate change and the policies to address it. Chapter 19, in particular, contains a great discussion of the economics of carbon tax and cap and trade.

See www.andrewdessler.com/chapter12 for additional resources for this chapter.

# Problems

1. a) Explain how a carbon tax works.
   b) Explain how a cap-and-trade system works.
   c) What is the fundamental difference between these two policies?
   d) Given a carbon tax of $x$ dollars (or a permit price of $x), an emitter will reduce emissions until what criterion is satisfied?
2. Why are emissions reductions achieved by use of a carbon tax or cap-and-trade system cheaper than those achieved by use of conventional regulations?
3. a) What is an offset?
   b) What does *additionality* mean?
4. In a *New York Times* op-ed piece (December 6, 2009), climate scientist Jim Hansen makes the following argument: "Consider the perverse effect cap and trade has on altruistic actions. Say you decide to buy a small, high-efficiency car. That reduces your emissions, but not your country's. Instead it allows somebody else to buy a bigger S.U.V – because the total emissions are set by the cap." He argues that this renders a cap-and-trade system ineffective. Why is this argument wrong?
5. Imagine a carbon tax is implemented. One day, you decide not to drive to the grocery store, and you apply for offset credit for the emissions that did not occur because this trip was not taken. Should you get paid for this? What would you have to prove in order to get paid?
6. For the following, assume that Plants A and B have the following marginal costs for reducing emissions:

| Number of units reduced | Marginal costs for Plant A | Marginal costs for Plant B |
|---|---|---|
| 1 | 3 | 1 |
| 2 | 5 | 2 |
| 3 | 7 | 3 |
| 4 | 9 | 5 |
| 5 | 11 | 9 |

   a) The government tells both plants to reduce three units of output. How much does this "conventional" regulation cost each plant? What is the total cost?
   b) The government implements a carbon tax of $5 per unit. How much does each plant reduce? What is the total cost?
   c) Which approach is cheaper? Why is the cheaper approach cheaper?

7. The table below shows the marginal costs of two plants, each of which emits 10 units each year. They both have six permits, meaning that if they do not trade, they each would have to reduce 4 units. Assume that they are the only two actors in the market, so the prices are set by their marginal costs.

| Number of units reduced | Marginal costs for Plant A | Marginal costs for Plant B |
|---|---|---|
| 1 | 1 | 3 |
| 2 | 2 | 6 |
| 3 | 3 | 9 |
| 4 | 4 | 12 |
| 5 | 5 | 15 |
| 6 | 6 | 18 |
| 7 | 7 | 21 |

   a) How many permits will Plant B buy from Plant A?
   b) In what price range will these permits exchange hands?

8. Why will voluntary and informational approaches not lead to deep reductions in emissions?

9. For a closer-to-home example of a cap-and-trade system, imagine the following scenario: Your professor gives everyone five points of extra credit on the final exam. Further, the professor says that you can sell your extra credit to other students. What would you do? Sell yours, buy extra, or just hold on to your five points? More generally, which students will sell their extra credit and which will buy more?

10. When you buy an airline ticket, you can also buy a "carbon offset" that will cancel out the emissions from the flight. They typically do not provide much information about the carbon offset. Under what conditions would it be a good thing to buy? Would you buy one?

11. Another cap-and-trade system is the Regional Greenhouse Gas Initiative, more commonly referred to by its initials, RGGI (and pronounced "reggie"). RGGI is made up of several Northeastern states running from Maine to Maryland. In a paragraph or two, explain the details of the RGGI and whether it is been a success.

12. One argument made by those who oppose reducing emissions is, "The energy sources we use are always the cheapest and most plentiful – which are coal, oil and natural gas. Wind, solar, etc. are more expensive and therefore bad for the economy." What is right and what is wrong about this argument?

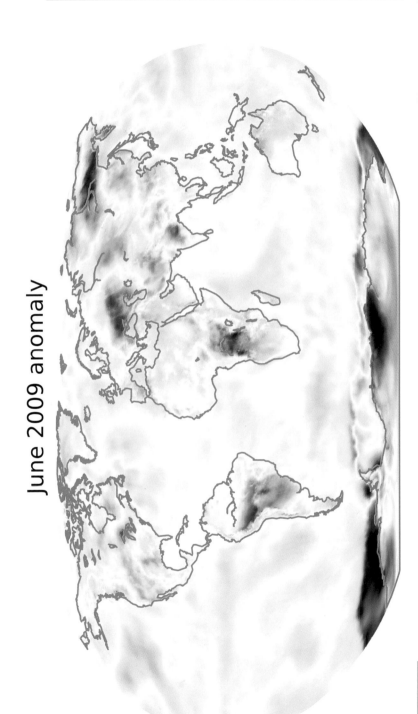

June 2009 anomaly

6.0
4.5
3.0
1.5
0.0
−1.5
−3.0
−4.5
−6.0

**Plate 2.1**   The monthly surface temperature anomaly in June 2009 in degrees Celsius. The reference temperature for the anomaly is the average of the June temperatures from 1979 to 2009 (data obtained from the MERRA reanalysis).

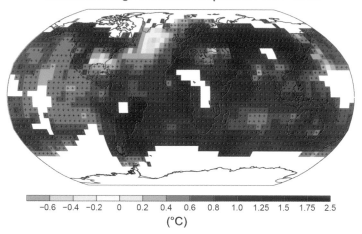

Observed change in surface temperature 1901–2012

-0.6 -0.4 -0.2 0 0.2 0.4 0.6 0.8 1.0 1.25 1.5 1.75 2.5
(°C)

**Plate 2.3** The distribution of warming (in °C) between 1901 and 2012. Regions where data are too sparse to produce an estimate are white. Adapted from Figure SPM.1 of IPCC [2013].

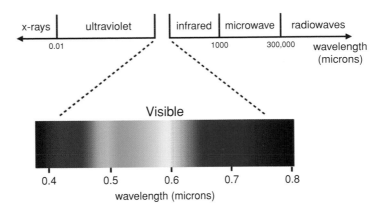

**Plate 3.1** The electromagnetic spectrum. Note that the visible part makes up only a minor part of this spectrum.

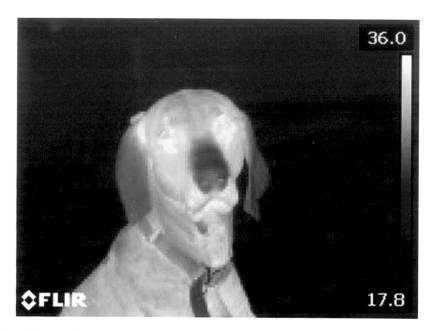

**Plate 3.6** Photo of Kasper Dessler in the infrared, with colors assigned to different temperatures.

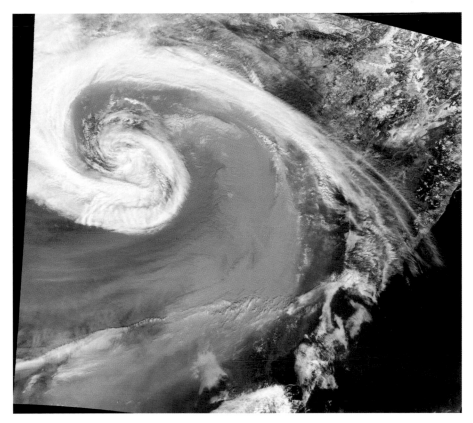

**Plate 6.5** Image of a strong temperate cyclone over China, pushing a wall of dust as it moved. The image was captured in early April of 2001 by the Moderate Resolution Imaging Spectroradiometer on NASA's Terra satellite (image obtained from the Earth Observatory; see earthobservatory.nasa.gov/IOTD/view.php?id=8341).

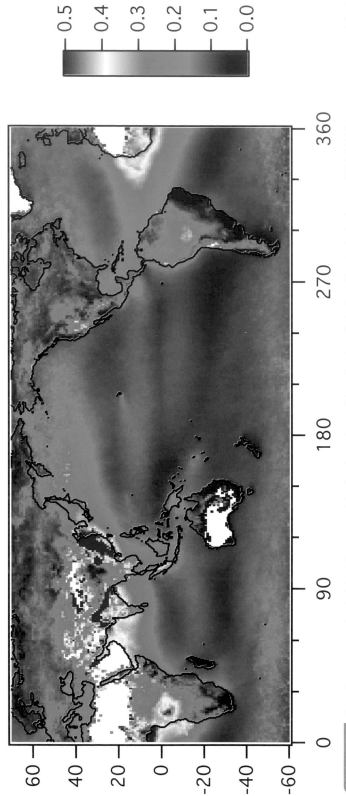

**Plate 6.7** Annual average aerosol optical depth (a measure of the abundance of aerosols) as a function of latitude and longitude, for the years 2004–2008 (measurements were made by the Moderate Resolution Imaging Spectroradiometer onboard NASA's Aqua satellite and were obtained from the NASA Goddard Earth Sciences Data and Information Services Center). White areas are regions where no data were obtained.

# Annual mean surface air temperature change

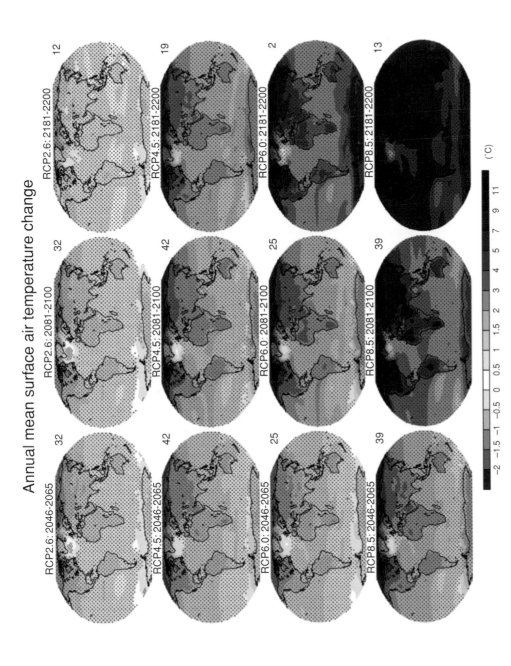

**Plate 9.1** The distribution of annual-average warming in the middle of the twenty-first century (left-hand column), the end of the twenty-first century (center column), and the end of the twenty-second century for the four RCP scenarios. Temperature increases are relative to the 1986–2005 average. These are calculated from an ensemble of climate models, with the number of models indicated in the upper right corner of each panel. Hatching indicates regions where there is disagreement among the models on the sign of the change, while stippling indicates regions where the models all agree on the sign of the change. This figure is adapted from figure 12.11 in Collins et al. [2013].

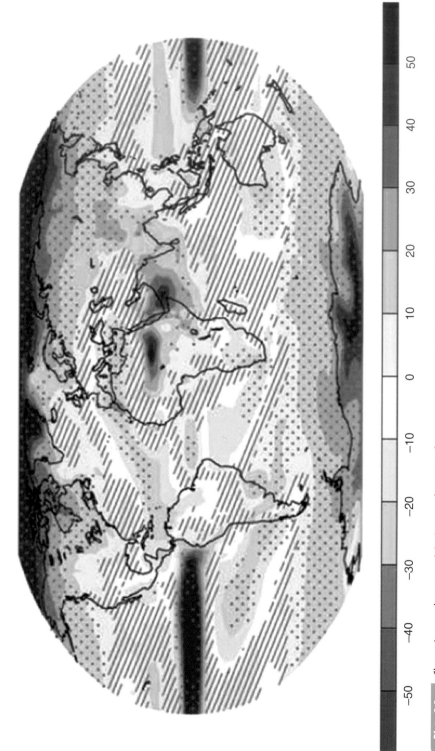

**Plate 9.2** Change in annual mean precipitation over the twenty-first century as predicted by climate models driven by the RCP8.5 high-emissions scenario. The change is the percent change of 2081−2100 precipitation relative to the 1986−2005 precipitation. This is the average of thirty-nine models' predictions. Hatching indicates regions where there is disagreement among the models on the sign of the change, while stippling indicates regions where the models all agree on the sign of the change. (adapted from IPCC 2013, figure SPM.8).

(%)

−50    −40    −30    −20    −10    0    10    20    30    40    50

# 13

# A brief history of climate science and politics

In Chapters 11 and 12, we explored our options for addressing climate change. We can adapt to the change, we can mitigate it by reducing the emissions of greenhouse gases, or we can geoengineer the climate. In Chapter 14, I will pull all these together so we can explore how we can choose among these options. Before we get to that discussion, though, I describe the context of the policy debate by providing a brief history of climate change science, policy, and politics.

## 13.1 The beginning of climate science

People have been speculating on the nature of the climate for millennia, but modern climate science began in earnest two centuries ago, in the early nineteenth century. In the 1820s, mathematician Joseph Fourier provided one of the first descriptions of the physics we now know as the greenhouse effect: A planet's atmosphere can trap heat and warm the surface of the planet beyond what it would be if it were a bare, airless rock (we covered this in Chapter 4). Several decades later, in 1859, physicist John Tyndall discovered that it was primarily water vapor and carbon dioxide in the atmosphere that provided the warmth, despite that fact that these two constituents make up just a small fraction of the atmosphere.

The first recognition that the climate could change occurred in the 1830s, when geologist Louis Agassiz and others identified glacial debris scattered across Europe. They correctly concluded that northern Europe must have previously been covered by ice. This was an unanticipated discovery; prior to that time, everyone had simply assumed that the climate they experienced was what it had always been and always would be. This discovery of widespread ice ages showed that climate had changed in the past, and it certainly suggested that it could change again in the future. This motivated much of the scientific study of climate over the next century.

By the end of the nineteenth century, our knowledge of the climate system was advancing rapidly. In 1896, scientist Svante Arrhenius, a Nobel Prize winner famous for his studies of chemical reactions, estimated the climate sensitivity – the warming of the planet from doubling the amount of carbon dioxide in the atmosphere – and found a value of 5 to 6°C. This is a bit higher than modern estimates of climate sensitivity of 1.5 to 4.5°C but still a remarkable achievement given what he knew about the planet.

Although Arrhenius' calculations were primarily focused on explaining the ice ages, he also realized that humans were adding carbon dioxide to the atmosphere from coal combustion. He estimated, however, that it would take thousands of years before humans would emit enough carbon dioxide to significantly warm the climate. He did not appreciate that fossil fuel use was growing exponentially. As we learned in Chapter 10, the growth of exponentials is so fast that it is easy to underestimate the long-term trajectory.

The work of Arrhenius really marks the beginning of the theory of human-induced global warming. But while the bare outlines of modern climate science were apparent at that time, many fundamental aspects of climate science were still not well understood. Whereas Arrhenius had suggested that the carbon dioxide emitted by humans would accumulate in the atmosphere, many scientists thought that most of the carbon dioxide emitted by humans would be quickly absorbed by the oceans (as discussed in Chapter 5, some carbon dioxide is indeed quickly absorbed by the ocean, but much is not). Furthermore, some scientists suggested that water so dominated the absorption of infrared radiation by the atmosphere that adding carbon dioxide would have no effect.

In addition, there was much less concern for environmental issues at that time. Nature was viewed as dangerous – in fairy tales, children who wandered into the woods did not come back. Indeed, the twists of weather and climate were among nature's cruelest weapons. When a tough winter or a severe drought could kill you, changing the climate does not seem like such a big deal.

So if the elements of human progress, such as the burning of fossil fuels, changed the climate, that was okay. Cutting down forest and replacing it with farmland or hunting predators like wolves to extinction were considered improvements. Nature was the enemy. Today, of course, we think differently about nature. We recognize that humans are strong enough to radically change our environment, and we therefore view the wilderness as something to be protected and conserved (although sometimes we do not act that way). "Nature" is somewhere you may go on vacation, if you can find it and afford to travel there. Figure 13.1 schematically illustrates the power shift between nature and humans over the past century.

As an example, consider the fairy tale of Little Red Riding Hood. The end of that story finds the woodsman killing the wolf with an ax and then cutting the wolf open and rescuing the grandmother. Unlike today, no one in the nineteenth century felt bad for the wolf. A modern version of that fairy tale would end quite differently: Everyone would realize that it was not the wolf's fault it ate grandma, so biologists from the U.S. Fish and Wildlife Service would dart and tranquilize the wolf, extract grandma without injuring the wolf, then radio tag the wolf and release him into a national park – where he would live happily ever after and (hopefully) refrain from further consumption of grandmothers.

Temperatures rose during the first few decades of the twentieth century (see Figure 2.2a), and by the 1930s it was apparent that the planet was warming. As *Time* magazine put it in 1939, "gaffers who claim that winters were harder when they were boys are quite right . . . weather men have no doubt that the world at least for the time being is growing warmer."[1]

Around this same time, Guy Stewart Callendar, an English inventor and engineer, suggested that this warming was caused by human emissions of carbon dioxide. This

---

[1]  Quoted in Weart (2008).

**Figure 13.1** Author's artistic impression of how people viewed their relationship with nature in the nineteenth century and today.

is likely the first time that someone suggested that human-induced climate change was underway. His work built off Arrhenius' observation that the burning of fossil fuels would warm the planet, but he revisited old measurements of atmospheric carbon dioxide and, unlike Arrhenius, realized that humans were already increasing the global atmospheric level of carbon dioxide. Like most other people of this time, however, Callendar was not terribly worried about any detrimental effects of human modification of the environment.

## 13.2 The emergence of environmentalism

By the 1950s, our view of the environment was changing as a result of several factors. One was the invention of nuclear weapons. Nuclear bombs with a yield of tens of kilotons had been used twice in World War II. By the 1950s, hydrogen-fusion bombs with yields hundreds of times larger had been developed. In fact, a single 1950s-era nuclear-armed bomber could carry bombs with more explosive power than all of the explosives used in World War II. It dawned on people that we humans now possessed the power to annihilate ourselves – and people suddenly realized that humans were now a power in many ways comparable to nature.

Air pollution was also becoming an important issue. Probably the most famous air-pollution event in history was the *killer smog of London* in 1952. In London at that time, most homes were heated with coal, which dumped sooty smoke into the London air. In early December 1952, a temperature inversion over London created a stagnant air mass over the city. As people burned coal, dark soot accumulated above the city. This dark cloud hung over London, blocking out sunlight, which caused the temperature to plummet. This caused people to burn more coal for heat, leading to even more soot in the air. During the height of the event on Sunday, December 7, the visibility in London was 1 foot. Cattle in the city's market were killed and their carcasses discarded rather than sold because their lungs were black. The particulates harmed people's health and killed many of the weak and old. On December 9, the weather changed and the killer fog was blown away, vanishing as quickly as it had arrived – but not before several thousand Londoners had died.

Air pollution in the United States was bad, too. My father was an undergraduate at Caltech, and he told me that visibility was so bad in Los Angeles when he arrived in the fall of 1948 that he was initially unaware that there were mountains nearby. After he would been there a few weeks, a rainstorm blew through and enormous mountains a few miles from campus were suddenly visible. He was, to put it mildly, surprised to see them. Such persistent air-pollution problems occurred throughout the world, and this further demonstrated that the human impacts on the environment were not always for the better.

At the same time, people in many parts of the world were getting richer. People who are poor tend to worry about where their next meal is coming from or where there are going to sleep tonight – they are not terribly concerned with the environment. However, as people become richer and have disposable income to spend on less essential things, protecting the environment becomes a higher priority. Once you have money, you care about your view of the mountains, where you are going to go camping this weekend, and the extinction of polar bears.

This increasing interest in the environment was bolstered by the International Geophysical Year, which took place in 1957 and 1958. This was an international effort that coordinated pole-to-pole observations of the Earth in order to improve our understanding of the fundamental geophysical processes that govern the environment. This intensive year of observations greatly improved our understanding of the Earth – and of the myriad of ways that humans can alter it. One of the most famous measurements started during the International Geophysical Year was of atmospheric carbon dioxide, also known as the Keeling curve (plotted in Figure 5.6). Within just a few years after commencement, these measurements showed that atmospheric carbon dioxide was rising as a result of human activities. Here was direct evidence of man's massive footprint on the planet.

## 13.2.1 The tobacco strategy

Around this same time, a seemingly unrelated debate was raging in our society over the heath effects of smoking. By the early 1960s, large longitudinal studies had proven that smoking cigarettes was associated with various health risks, such as an increased risk of lung cancer. In 1964, U.S. Surgeon General Luther Terry issued a report entitled *Smoking and Health*,[2] which detailed the dire health consequences of smoking. The evidence supporting their conclusions was enormous – the report

---

[2] profiles.nlm.nih.gov/NN/B/B/M/Q/

summarized nearly 7,000 scientific articles relating smoking and disease. The report made newspaper headlines across the country and was the lead story on television newscasts.

In response, the tobacco companies developed what has become colloquially known as the *tobacco strategy* – a concerted effort to cast doubt on established science in order to advance a particular policy goal. The goal of the tobacco strategy was not to prove that cigarettes were safe but rather to create doubt, as described in a tobacco company document from 1969:

> Doubt is our product since it is the best means of competing with the "body of fact" that exists in the mind of the general public. It is also the means of establishing a controversy.[3]

In attempting to generate doubt, the tobacco companies developed a set of actions to advance their strategy. This included:

- Finding a small number of sympathetic scientists who would convey the message of doubt to the general public. Misrepresent this to suggest that there is a vigorous debate in the scientific community.
- Cherry-pick data and focus on a small number of unexplained or anomalous details. Ignore the fact that the vast, vast, vast majority of data solidly supports the consensus view.
- Create the impression of controversy simply by asking questions, even if the answers were known and did not support the tobacco companies' case.
- Under the guise of fairness, demand equal time from media outlets to present tobacco companies' side.

Following this strategy, tobacco companies were able to keep the public debate over the health impacts of smoking alive for decades after it was settled in the scientific community. This episode would be a sad historical footnote if not for the fact that the tobacco playbook has been used again and again to cast doubt on science suggesting that humans are harming our environment.

## 13.3 The 1970s and 1980s: Supersonic airliners, acid rain, and ozone depletion

Environmentalism may have begun in the 1950s, but several events in the 1970s solidified it in the general public's consciousness. Up first was the battle over the development of a supersonic airliner. Given the increase in speed of aircraft from the original Wright Flyer through jet airliners like the Boeing 707, it was generally believed in the 1960s and 1970s that the next step for commercial aviation was a supersonic airliner. And the country that developed the first successful supersonic

---

[3] A large number of tobacco company documents can be viewed on the Legacy Tobacco Documents Library (see legacy.library.ucsf.edu/). This particular document can be found at legacy.library.ucsf.edu/tid/wjh13f00/pdf.

airliner would garner enormous national pride as well as economic benefits. Because of this, Europe, the United States, and the Soviet Union engaged in a three-way race to develop such an airliner in a competition much like the race to the moon.

In the early 1970s, however, scientists realized that a fleet of supersonic airliners might have serious environmental consequences. Jet engine exhaust includes chemicals that can destroy ozone, and because supersonic airliners fly at high altitudes for efficiency, these effluents would be dumped directly into the stratospheric ozone layer. Scientists began to worry that, given a large enough fleet of these planes, a significant loss of ozone might result. Because stratospheric ozone blocks high-energy ultraviolet photons, which harm plants and cause skin cancer in humans, the loss of this ozone could have serious detrimental effects on the biosphere.

In the end, the Europeans developed their supersonic airliner, the Concorde, while the Soviet Union developed their version, the Tu-144. However, fewer than two dozen Concordes and Tu-144s were ever built. The Concordes flew for three decades, while the Tu-144s flew only a handful of commercial flights before they were removed from service. The United States completely abandoned its efforts to build a supersonic airliner. Although the concern over ozone depletion did play role in limiting the success of these supersonic planes, it was mainly economics: supersonic planes are expensive, and consumers would rather have cheap tickets than fast planes. In fact, today's modern planes, such as the Boeing 777, actually fly slower than a Boeing 707 – but they trade speed for lower operating costs, allowing airlines to offer lower prices. As a result, fares in 2005 for flights from New York to Los Angeles – roughly $300 round trip – were about the same in nominal dollars as they were in the 1940s.

In 1973 and 1974, just as the supersonic transport debate was subsiding, scientists first theorized that man-made halocarbons might deplete ozone. The issue of ozone depletion had already entered the public sphere during the supersonic airliner debate, so policymakers and the general public were familiar with this risk. At this time, the threat from halocarbons (frequently referred to as CFCs, which stand for *chlorofluorocarbons*) was completely theoretical – it would be more than a decade before actual observations of ozone depletion were obtained – and there was no effective replacement for CFCs in many applications. Nevertheless, in the late 1970s, the United States banned CFCs from being used in some nonessential applications – such as a propellant in aerosol spray cans.

In response, CFC manufacturers and industries that used CFCs in their products joined together to defend the molecule. To do so, they took the tobacco strategy – attack the science! – and derived a new version optimized for environmental issues. They focused on three main claims: It is not happening; if it is happening, we are not to blame; and, if we are to blame, then fixing the problem will be too expensive. As with the tobacco debate, the goal of those defending CFCs was not to prove any of these points, but to simply generate enough uncertainty and doubt to stall any policy to regulate CFCs.

At this same time, another environmental problem was emerging: acid rain. Many power plants, particularly those that burn coal, emit large amounts of sulfur dioxide and nitrogen oxides to the atmosphere. Once in the atmosphere, these molecules can be absorbed by cloud droplets and raindrops, and once dissolved in water they react to form sulfuric acid and nitric acid. This is analogous to the way carbon dioxide dissolves into water to form carbonic acid (discussed in Chapter 5). The difference is that carbonic acid is what is known as a weak acid, whereas nitric and sulfuric acids are strong acids. This means that dissolving nitrogen oxides or sulfur dioxide into water makes the resulting liquid much more acidic than dissolving an equal number of molecules of carbon dioxide. When this acid rain falls to the ground, the types of potential damage it can do are numerous: bleaching of nutrients from soils, acidification of lakes and rivers, damage to wildlife and plants, damage to human-built structures, and so on.

This entire theory of acid rain is scientifically quite simple, and research done over the 1970s and 1980s definitively connected emissions from power plants to the acidic precipitation. In response to this research, the first broad international agreement covering acid rain, The Convention on Long Range Transboundary Air Pollution, was signed in Geneva by thirty-four member countries of the U.N. Economic Commission for Europe on November 16, 1979. The next year, the Council of the European Communities enacted a directive reducing sulfur dioxide emissions.

The Reagan Administration, however, was resistant to enacting regulations on emissions of acid-rain precursors. To support their reticence, the administration created a series of reports that argued that there was too much uncertainty to take action:

> "The state of the science . . . probably will not yield a scientifically complete assessment of acid deposition in the next few years," says the report prepared for Congress and the Reagan Administration. "To date, the state of the science will not allow assertive recommendations. Trends are weak and evasive. Data are spotty. One of the most basic uncertainties is the extent of damage caused by acid deposition and its rate of change." The report says air pollutants, some of which cause acid rain, are the prime suspects in investigation of dying forests in the northeastern United States but "the association between damage and the occurrence of those pollutants is not well defined." It says some lakes have been acidified and fish have died but "the rate, character and full extent of these changes remain major scientific unknowns." And it says acidity can accelerate corrosion of buildings but "information about effects on materials from acid deposition needs to be better defined."[4]

The United States took no action on acid rain during the Reagan Administration. However, in the early 1990s, the George H. W. Bush Administration enacted regulations to reduce sulfur emissions through a cap-and-trade system – just like the cap-and-trade systems discussed in Chapter 12. These regulations greatly reduced the emissions of sulfur at a price far below expectation. And, as expected, this greatly decreased the occurrence of acid rain. The success of cap and trade in helping solve the acid rain problem is one of the main reasons that policymakers looked hopefully at that mechanism for addressing climate change.

It is worth reiterating that the cap-and-trade solution in the United States to the acid rain problem emerged from the Republican George H. W. Bush Administration. At the time, it was viewed as a conservative-friendly method of solving the problem because it let the market determine how emissions reductions would be allocated. For this same reason, environmentalists were suspicious or outright opposed to cap and

---

[4] "Acid Rain Facts Called Sketchy," *The Globe and Mail* (Canada), June 12, 1984.

trade. It was unclear if it would actually reduce acid rain, and many environmentalists viewed the program as allowing emitters to pay to pollute. If pollution was wrong, they reasoned, you should not be able to pay to do it. Over time, and with the incredible success of the program, environmentalists came around to viewing cap and trade as a particularly effective way to solve environmental problems. This is why they started thinking of solving the climate change problem using this same policy instrument. At the same time, Republicans have developed a steadfast opposition to these cap-and-trade schemes.

The ozone problem remained an active scientific and political issue throughout the early and mid-1980s. The original theories from the 1970s suggested that ozone depletion would be a slow process, taking half a century or longer for significant depletion of ozone to occur, and that it would primarily occur at mid-latitudes and high altitudes. But when scientists obtained the first evidence that ozone was actually being depleted as a result of CFCs, they found it was occurring much more rapidly and in a different place than predicted. They observed extremely rapid loss of ozone over Antarctica, where every spring, roughly 90 percent of the ozone in the lower stratosphere was being destroyed in a month or so (it built back up during the rest of the year so that it was available to be destroyed again the following year). This annual loss of ozone became known as the *ozone hole.*

Within a few years, newly discovered chemical reactions that relied on CFCs combined with the unique meteorology of the polar regions were identified as the cause of this rapid polar ozone depletion. This confirmed the role of humans and suggested that this problem might be more serious than had previously been recognized. In response to this threat, the world adopted the *Montreal Protocol* in 1988, an international agreement committing the world to phasing out CFCs.

An important aspect of the Montreal Protocol was that the phase-out of CFCs happened in two stages. Industrialized countries phased out CFCs first, followed ten years later by developing countries. There are several reasons for this. Industrialized countries are richer than the developing countries, so they have more resources to apply to phasing out CFCs. Moreover, by having the rich countries go first, economies of scale and technical advances would be expected to drive down the cost for developing countries of phasing out CFCs. There were also ethical considerations. The CFCs in the atmosphere – which were causing the ozone depletion – had mainly been released to the atmosphere by activities in the industrialized countries. Developing countries had contributed little to the problem. Thus, it was agreed that industrialized countries should take the first step to clean up the problem.

During and after the negotiation of the Montreal Protocol, the science of ozone depletion was advancing rapidly, and evidence continued to accrue about the dangers CFCs posed to the ozone layer. But even as the science became more certain, so-called ozone skeptics stepped up their attacks on the science of ozone depletion. A good example is this 1989 quote from *National Review:*

> The current situation can fairly be summarized as follows: The CFC-ozone theory is quite incomplete and cannot as yet be relied on to make predictions. The natural sources of stratospheric ozone have not yet been delineated, theoretically or experimentally. The Antarctic ozone hole is ephemeral; it comes and goes, and seems to be controlled by climatic factors outside of human control rather than by CFCs.

The reported decline in global ozone may be an artifact of the analysis. Even if real, its cause may be related to the declining strength of solar activity rather than to CFCs. The steady increase in malignant melanoma has been going for at least 50 years and has nothing to do with ozone or CFCs. And the incidence of ordinary skin tumors has been greatly overstated. . . . And substituting for CFCs is no simple matter. A New York Times report of March 7, 1989 talks about the disadvantages of the CFC substitutes. They may be toxic, flammable, and corrosive; and they certainly won't work as well. They'll reduce the energy efficiency of appliances such as refrigerators, and they'll deteriorate, requiring frequent replacement. Nor is this all; about $135 billion of equipment use CFCs in the U.S. alone, and much of this equipment will have to be replaced or modified to work well with the CFC substitutes. Eventually that will involve 100 million home refrigerators, the air-conditioners in 90 million cars, and the central air-conditioning plants in 100,000 large buildings. Good luck! The total costs haven't really been added up yet.[5]

If this argument sounds familiar, it is because it is the tobacco strategy. And in retrospect, we now know that all of these arguments are wrong. Two decades of research have concretely verified the link between CFCs and stratospheric ozone depletion. What is more, the costs of replacing CFCs with ozone-safe alternatives turned out to be so small that, when CFCs were completely phased out in the mid-1990s, virtually no one noticed.

In addition to the scientific uncertainty arguments, the ozone skeptics also provided a context for why so many scientists, politicians, and activists were convinced by such shoddy science:

It's not difficult to understand some of the motivations. For scientists: recognition for keeping dusty records or running complicated computer models that are rather dull; more grants for research; press conferences; and newspaper stories. Also the feeling that maybe they are saving the world for future generations. For bureaucrats the rewards are obvious. For diplomats there are negotiations, initialing of agreements, and – the ultimate – ratification of treaties. It doesn't really much matter what the treaty is about, but it helps if it supports "good things." For all involved there is of course travel to pleasant places, good hotels, international fellowship, and more. It's certainly not a zero sum game.

I have left environmental activists to the last. There are well-intentioned individuals who are sincerely concerned about what they perceive as a critical danger to the health of future generations. Many of the professionals share the same incentives as government bureaucrats: status, salaries, perks and power. And then there are probably those with hidden agenda of their own – not just to "save the environment" but to change our economic system. The telltale signs are the attack on free enterprise, the corporation, the profit motive, the new technologies. Some are socialists, some are Luddites.

Most of these "compulsive utopians" have a great desire to regulate – on as large a scale as possible. To them global regulation is the "holy grail." That's what makes the CFC-ozone issue so attractive to them.

Thus, an alternative narrative is created: The science the public is hearing in the mainstream media is wrong, and the reason wrong science is being conveyed is that

---

[5] S. F. Singer, "My Adventures in the Ozone Layer," *National Review,* June 1989.

scientists and advocates are corrupt, biased, or stupid. No evidence is provided to support these charges, but it is not really necessary. The point here is to cast doubt, not to prove the accuracy of this narrative.

The Earth's temperature remained relatively constant between the 1940s and 1970s (Figure 2.2). Despite this, research since the 1950s had emphasized the risk of global warming due to increasing abundances of atmospheric greenhouse gases. At the same time, however, the abundance of aerosols from human activities was also rising (e.g., from the burning of high-sulfur coal). As discussed in Chapter 6, aerosols can cool the planet, which offsets some of the warming from increased greenhouse gases. Some scientists suggested that humans were in fact adding enough aerosols to the atmosphere to overpower greenhouse gases and that the dominant influence of man was a net cooling of the climate.

A legitimate scientific debate ensued over which effect would dominate, and by the end of the 1970s the debate had been settled in favor of those predicting that global warming would be the dominant human influence. In recent years, some have misrepresented this mere existence of a debate to suggest that the scientific community in the 1970s was predicting global cooling. This is incorrect – there was never any widespread consensus among scientists that aerosol-induced cooling was the dominant influence of humans.[6]

The 1970s ended with the publication of an influential report[7] from the U.S. National Academy of Sciences that reviewed the science and came to this conclusion: "If carbon dioxide continues to increase, the study group finds no reason to doubt that climate changes will result and no reason to believe that these changes will be negligible. The conclusions of prior studies have been generally reaffirmed." More quantitatively, they concluded that a doubling of carbon dioxide would result in a warming of 1.5 to 4.5 °C, which is the same as today's estimates. More research in the early 1980s fleshed out and confirmed the general view that humans were in the process of modifying the climate. However, the general public and most politicians were not yet focused on the issue.

## 13.4 The year everything changed: 1988

1988 was the year when climate change went from being a mostly academic problem to a political one. The United States was blisteringly hot that summer, with much of the country suffering drought conditions and temperature records smashed on a seemingly daily basis. A small number of U.S. congressional leaders were interested in the problem of climate change, and they felt the time was right to hold a congressional hearing on it. In a stroke of political genius, they held the hearing in August, which is the hottest part of the Washington summer. Committee staffers opened the windows

---

[6] For a good review of the history of "global cooling" and how it is misrepresented in today's debate, see Peterson et al., "The Myth of the 1970s Global Cooling Scientific Consensus," *Bulletin of the American Meteorological Society* 89 (2008): 1325–1337.

[7] Ad Hoc Study Group on Carbon Dioxide and Climate, *Carbon Dioxide and Climate: A Scientific Assessment* (Washington, DC: Climate Research Board, National Research Council, 1979).

of the hearing room the night before, so the room itself was also extremely hot. Videos of the hearing show people sweating and mopping their brow, effectively reinforcing the message about climate change.

At that hearing, NASA climate scientist James Hansen declared that he was 99 percent confident that the world really was getting warmer and that there was a high degree of probability that it was due to human activities. Coming on the heels of the publicity over the ozone hole, this created a media firestorm, and it put the issue of climate change onto the political radar. In the next few months, the United Nations passed a resolution urging the "Protection of global climate for present and future generations of mankind." *Time* magazine, instead of naming a "Person of the Year" for 1988, named "Endangered Earth" the "Planet of the Year."

This was also the year that the Intergovernmental Panel on Climate Change, or IPCC, was formed (this organization was discussed in Chapter 1). During the negotiation of the Montreal Protocol, the World Meteorological Organization and the U.N. Environmental Program had put out a series of reports that described the science of stratospheric ozone depletion. These assessments were incredibly successful at establishing the bedrock scientific principles upon which the Montreal Protocol was based. The climate policy community, seeing the success of these ozone assessments, concluded that similar assessments about climate science would help facilitate those policy debates, and this was the job the IPCC was created to do.

As momentum for enacting regulations to reduce emissions began to grow, so did the pushback from those opposed to regulations. Given that the energy is a several trillion dollar per year business, it should come as little surprise that many people and institutions strenuously opposed regulations that would cost them some of these trillions.

They were joined by those philosophically opposed to environmental regulations. This group was mainly motivated by the fundamental belief that regulations on greenhouse gas emissions were an unacceptable infringement on freedom. This is summed nicely up by Vaclav Klaus, President of the Czech Republic (and one of the very few leaders of any country to doubt the mainstream view of the science of climate change): "The largest threat to freedom, democracy, the market economy, and prosperity at the end of the twentieth and at the beginning of the twenty-first century is no longer socialism. It is, instead, the ambitious, arrogant, unscrupulous ideology of environmentalism."[8]

With the tobacco strategy in mind, those opposed to regulations on greenhouse gases focused on attacking the science. To do this, they recruited a small group of contrarian scientists to make the public argument. Many of these so-called *climate skeptics* were veterans of previous battles – tobacco, acid rain, ozone – and they had deep experience casting doubt. The skeptics' views of science are heterogeneous; for example, some skeptics dispute that the Earth is warming, while others accept that the Earth is warming but dispute that humans are responsible. Some even dismiss the fundamental physics of the greenhouse effect. But they typically share a disdain for any science that might lead to increased government regulation.

---

[8] V. Klaus, *Blue Planet in Green Shackles* (Washington, DC: Competitive Enterprise Institute, 2007).

In 1990, the IPCC put out its first assessment on the science of climate change. In it, the IPCC concluded that "the size of this [observed] warming is broadly consistent with predictions of climate models, but it is also of the same magnitude as natural climate variability. Thus the observed increase could be largely due to this natural variability." This relatively weak statement about the role of humans in climate change reflected legitimate uncertainties in climate science at that time. Because of these uncertainties, a definitive attribution of the warming to greenhouse gases was not possible.

## 13.5 The Framework Convention on Climate Change: The First Climate Treaty

Despite this, many world leaders felt that action had to be taken on climate change. The result was the Earth Summit in Rio de Janeiro in 1992, from which emerged the treaty known as the *Framework Convention on Climate Change,* frequently referred to simply as the "framework convention" or by its initials, FCCC. The FCCC enjoys near-universal membership, with 192 countries having ratified it, including the United States, China, and all other big emitters. The principles enshrined in the FCCC remain the major building blocks on which negotiations of treaties to reduce emissions have been built.

The most contentious debate over climate change policies involves mitigation. In that regard, the stated goal of the FCCC is "to achieve stabilization of greenhouse gas concentrations in the atmosphere at a low enough level to prevent dangerous anthropogenic interference with the climate system." This is fine as far as it goes, and it receives widespread agreement at this level of abstraction. In practice, though, the meaning of this statement hinges on the definition of the word *dangerous.* There is no scientific definition of what dangerous climate change is because this is a value judgment – climate change that one person may perceive as dangerous may not be perceived that way by someone else.

In order to bring fairness or *equity* to any climate change agreement, the FCCC also enshrines the concept of *common but differentiated responsibilities.* This means that all countries must participate in solving the climate change problem, but not necessarily the same way. For example, we might expect rich, industrialized countries to begin cutting their emissions first, with developing nations cutting their emissions later. The reasons for this are similar to the reasons that, in the Montreal Protocol, industrialized countries phased out CFCs first, followed ten years later by developing countries. The industrialized countries of the world are far richer than the developing countries, so they have more resources to apply to reducing emissions. Moreover, by having the rich countries go first, economies of scale and technological advancement would bring down the cost of reducing emissions so that, when developing countries did begin reducing their emissions, the cost to them would be less.

There are also moral considerations. The 2 billion or so poorest people in the world currently live hard lives of crushing poverty. One of the ways to raise these people out of poverty is through economic growth – increasing their consumption of goods and services. This requires energy, so anything that makes consuming energy harder or more expensive for the poorest will also make it harder to lift these people out of poverty. Common but differentiated responsibility is a way of saying that solutions to climate change should not work at cross-purposes to efforts to reduce poverty.

There is also the question of historical responsibility. Most of the increase in carbon dioxide in the atmosphere over the past 250 years is due to emissions from the rich and industrialized countries. In fact, the world's rich countries are rich because of the energy they consumed – and the emissions that resulted. Thus, it makes sense for them to have a greater responsibility for taking the first steps toward cleaning up the problem. It is also clear, however, that developing countries must eventually contribute. China is now the largest emitter of carbon dioxide, and several other developing countries are either presently major emitters or on track to be. Reducing global emissions significantly over the coming century will be impossible without these developing countries eventually making deep emissions reductions.

The FCCC also included what is referred to as the *precautionary principle*: "Where there are threats of serious or irreversible damage, lack of full scientific certainty should not be used as a reason for postponing such measures." This is a crucially important and widely misunderstood statement about the implications of scientific uncertainty in policy deliberations. It does NOT say that, if the risks are high enough, we must take action, regardless of scientific uncertainty. Rather, it says that, if the impacts of a risk are sufficiently serious, scientific uncertainty should not be used as an excuse to do nothing. There may be many other reasons to do nothing (e.g., economic, moral), but science should not be one.

The FCCC was intended to be a starting point for more specific and binding measures to be negotiated later. Consequently, in contrast to its ambitious principles and objectives, the treaty's concrete measures were weak. Under the FCCC, nations committed to reporting their current and projected emissions and supporting climate research. Parties also accepted a general obligation to adopt, and report on, measures to limit emissions. What these measures had to be, or had to achieve, was not specified. Only for the industrialized countries did this general obligation also include the specific aim of returning emissions to 1990 levels by 2000. This target was nonbinding, meaning that it was an aspirational target and there were no sanctions for missing the target.

## 13.6 The Kyoto Protocol

By the middle of the 1990s, it was clear that no country would achieve the emissions-reduction target set in the FCCC and that a treaty with mandatory reductions would be required. Around that same time, the IPCC released its second assessment of the science of climate change, in which it concluded that "the balance of evidence suggests a discernible human influence on the climate." This quote reflected the fact that the science of climate change had significantly advanced since the IPCC's first assessment and there was now much more evidence linking the observed warming to human activities. But significant uncertainties remained.

In response to these developments, the *Kyoto Protocol* was negotiated in December 1997. Unlike the FCCC's nonbinding emissions reductions, the Kyoto Protocol required emissions from participating industrialized countries, averaged over a commitment period running from 2008 through 2012, to be approximately 5 percent below their 1990 emissions level. Developing countries, however, had no emissions reduction requirements.

The Protocol incorporated several provisions to allow flexibility in how nations met their emission limits. As I discussed in Chapter 12, flexibility mechanisms allow emissions reductions to be shifted to where they can be made most cheaply, thereby reducing the overall cost of attaining a particular emissions target. Flexibility mechanisms included targets being defined for total emissions of a basket of carbon dioxide and five other greenhouse gases (methane, nitrous oxide, hydrofluorocarbons, perfluorocarbons, and sulfur hexafluoride). Countries could meet their target by reducing emissions of any of these gases, not just carbon dioxide.

Emissions-reduction obligations could also be exchanged between nations through various mechanisms. Under one mechanism, known as the Clean Development Mechanism, industrialized countries could invest in emissions reduction projects where it is cheapest globally – typically in developing countries – and count those reductions toward their own goal. For example, France can invest in a wind farm in China and count the reduction in emissions toward France's emissions target. Like offsets, such projects must satisfy additionality: it must be demonstrated that the wind farm in China would not have been built without the financial support of France through this program.

The Protocol also included provisions for nations to meet some of their obligation through offsets. The Protocol included credit for reforestation, but some countries, such as the United States, wanted credit for other carbon-capturing activities, such as agriculture. This turned into a significant conflict at a negotiating session in November 2000 in The Hague. A proposed compromise was almost reached, but it was ultimately rejected at the last minute by the French and German environment ministers, who judged that the proposed offsets weakened the Kyoto commitments too much.

## 13.7 The George W. Bush years: 2001–2008

Shortly after the breakdown in negotiations at the meeting in The Hague and just a few months after taking office in 2001, the Bush Administration announced it was withdrawing the United States from the Kyoto Protocol process. The reasons included too much scientific uncertainty about climate change and potential harm to the U.S. economy. Although it later retreated from claiming that its withdrawal was based on scientific uncertainty, the Bush Administration continued to hold that the Protocol was unacceptable because of the high costs to the U.S. economy.

A particular problem cited by the Bush Administration was the absence of emission limits for developing countries. Although "common but differentiated responsibilities" is enshrined in the FCCC – which George H. W. Bush signed – and had also been incorporated into the Montreal Protocol, the George W. Bush Administration painted this as unfair to the United States. While it is true that China was already a major economic competitor to the United States in many areas, China was also a much poorer country than the United States and had far fewer resources to devote to reducing emissions.

Also in 2001, the IPCC released its third assessment report on the science of climate change. The report came to this conclusion: "There is new and stronger evidence that most of the warming observed over the last 50 years is likely attributable to human activities." Compared with the previous reports, this one made a much more definitive statement about the role of humans in the recent warming. However, there was still uncertainty, as reflected by the use of the word *likely*, which denotes a two out of three chance.

In February 2002, President Bush outlined his alternative approach to the issue. While consistently saying that climate change was a problem that needed to be addressed, the Bush Administration steadfastly avoided talking about reducing greenhouse-gas emissions. Instead, the emphasis was on reducing the greenhouse-gas intensity – the $T$ term of the IPAT relation, which concerns emissions per dollar of GDP. Their stated goal was to reduce greenhouse-gas intensity by 18 percent by 2012. This was a weak goal because greenhouse-gas intensity has historically declined at 1 to 2 percent/year without any policies. Thus, the Bush goal would likely be met with little or no effort.

The Bush policies also increased funding for climate change science and for specific technologies to reduce emissions, implemented tax incentives for renewable energy and high-efficiency vehicles, and started several programs to encourage voluntary emission cuts by businesses.

While the Bush Administration made no serious effort to reduce emissions, the rest of the industrialized world continued pushing for the Kyoto Protocol. To enter into force – and so become binding on those who ratified – the Protocol required ratifications by fifty-five countries, including nations contributing at least 55 percent of 1990 industrialized-country emissions. This threshold meant that, after the withdrawal of the United States, the treaty could enter into force only if all other major industrialized countries joined. The fate of the Protocol remained uncertain until November 2004 when, after several years of uncertainty about its intentions, Russia submitted its ratification, allowing the Protocol to enter into force on February 16, 2005.

But the long delay awaiting the required ratifications meant that the Protocol entered into force only three years before the start of the five-year commitment period (2008–2012). Some nations, such as the European Union, took aggressive action by enacting a large-scale cap-and-trade program, the European Trading System (ETS) (this was discussed in detail in the previous chapter). Other countries made little effort to achieve what would be a large deflection of emissions over very few years.

At the same time, in the absence of any federal efforts to reduce emissions in the United States, efforts trickled down to the state and local levels. For example, Connecticut, Delaware, Maine, Maryland, Massachusetts, New Hampshire, New Jersey, New York, Rhode Island, and Vermont banded together to form the Regional Greenhouse Gas Initiative, more commonly referred to by its initials, RGGI (and pronounced "reggie").[9] The RGGI is a regional cap-and-trade program that covers emissions from just one economic sector – electric power plants. The Western Climate Initiative was formed by a group of western states and Canadian provinces to also develop a regional cap-and-trade system. In addition, many individual U.S. states and cities began efforts to reduce emissions.

In 2007, the IPCC released its fourth assessment on the science of climate change and came to this conclusion: "Most of the observed increase in globally averaged temperatures since the mid-twentieth century is very likely due to the observed increase in anthropogenic greenhouse gas concentrations." This continued the trend toward stronger statements implicating humans in the warming – the words *very likely* here denote 90 percent confidence.

With the end of the Kyoto Protocol in 2012 in sight, representatives of the world's governments met in Bali in December 2007 to begin negotiations for a new climate treaty that would build on what the Kyoto Protocol had accomplished, to be ready before a meeting in Copenhagen in 2009. Importantly, it was agreed in Bali that this new agreement would, unlike the Kyoto Protocol, include emissions reductions by both industrialized and developing countries. Subsequent negotiations quickly split over the relative efforts required of these two groups.

In 2008, as George W. Bush's presidency was reaching its end, the campaign to succeed him heated up. Senator Barack Obama was the Democratic nominee, and he was opposed by the Republican nominee, Senator John McCain. Both candidates accepted the reality of climate change and the need to do something about it – in fact, Senator McCain had tried several times over the previous decade to get an emissions reduction bill through the U.S. Senate. The disagreements between the candidates on climate policy were quite minor – e.g., arguments over how much we should rely on nuclear power to reduce emissions, how deep the cuts should be in 2050.

## 13.8 The Obama years: 2009–today

Barack Obama became president of the United States in January 2009. Shortly thereafter, in December 2009, an international meeting in Copenhagen took place to negotiate the follow-on to the Kyoto Protocol. While hopes were raised by renewed U.S. engagement under the Obama Administration, the Copenhagen meeting was marked by continuing disputes between developing and industrialized countries over sharing the burden of action. Developing nations wanted the industrialized world to make sharp, near-term (e.g., by 2020) reductions in emissions, whereas the industrialized world wanted the developing nations to agree to quantitative emissions reductions.

On the final day of the conference, President Obama and a handful of key developing country leaders negotiated an agreement known as the *Copenhagen Accord*. The Accord included these major points:

- Global temperatures should not rise more than 2°C beyond preindustrial temperatures.

---

[9]  In May 2011, New Jersey announced its intention to leave the program.

- Deep cuts in emissions will be necessary. Recognizing equity issues, these deep cuts may be delayed in developing and poor countries.
- The world's rich industrialized countries each agreed to set their own target for emissions in 2020. Since the adoption of the Accord, the European Union, for example, agreed to a reduction of 20 to 30 percent below 1990 emissions, while the United States agreed to a reduction of 17 percent below 2005 emissions.
- The world's developing countries agreed to take on mitigation efforts but did not accept specific emissions targets. China, for example, agreed to reduce greenhouse gas intensity (the T term in the IPAT relation) by 40 to 45 percent below 2005 levels by 2020. However, given the rapid economic growth of China, this still corresponds to increased emissions.
- Flexibility should be incorporated into policies in order to achieve emissions reductions at the lowest cost.
- Adaptation must necessarily be part of our response. Industrialized countries agree to provide resources to help poorer countries adapt.

Around that same time, Obama and his congressional allies began advancing major bills on health care and climate change through Congress. In response to this, as well as general opposition to President Obama's policies, the Tea Party, a libertarian wing of the Republican Party dedicated to reducing the role of government in our lives, became a major force in U.S. politics. Because policies to reduce emissions require some government intervention in the energy market – usually by pricing carbon emissions either through a carbon tax or a cap-and-trade system – the Tea Party rabidly opposes climate legislation.

The rise of the Tea Party put immense pressure on Republican politicians to reject climate change science and any legislation to reduce emissions. Before 2009, a number of prominent Republicans openly acknowledged the risk of climate change and supported policies to reduce greenhouse-gas emissions. This included leaders in the party such as John McCain (2008 presidential nominee), Mitt Romney (2012 presidential nominee), and Newt Gingrich (former speaker of the house). After 2009, however, each of these politicians adopted a skeptical position on the science of climate change and opposed legislation to reduce emissions. Republican politicians who did not adopt this viewpoint quickly found themselves out of a job.

In 2010, Tea Party-affiliated candidates, virtually all of whom reject the science of climate change or the seriousness of the problem, were elected in numbers high enough to fundamentally change the composition of Congress. This ended any opportunity to get comprehensive climate legislation through the U.S. Congress.

Conservative pushback on climate policy was also occurring in other countries. In Canada, the conservative government of Prime Minister Stephen Harper withdrew from the Kyoto Protocol in 2011. The government did this because of well-worn criticisms of the Protocol (e.g., it does not include developing countries) and because Canada has immense oil reserves in the form of tar sands, which will likely be worth significantly less if the world agrees to stringent emissions reductions. In 2013, a conservative government in Australia began rolling back a price on carbon emissions that had been implemented several years earlier by the previous government.

By the end of 2012, the end of the Kyoto Protocol's commitment period, most industrialized countries had not achieved their Kyoto Protocol targets (2008–2012 emissions about 5 percent below 1990 levels). Those that did, mostly in Central and Eastern Europe, relied on the fact that the base year was 1990, prior to the collapse of the Soviet Union. After the dissolution of the Soviet Union in 1991, much of the inefficient industry there was shut down, leading to a huge decrease in emissions; despite subsequent growth, emissions there had not yet climbed back to 1990 levels by the Kyoto Protocol's commitment period.

In 2012, Barack Obama was reelected president, and he made climate change a key issue in his second term. Given the composition of Congress, though, getting any climate legislation through it was clearly impossible. The Obama Administration instead turned to executive orders – policies the president can implement without congressional approval – and existing authority granted to the administration under the U.S. Clean Air Act to address climate change.

In 2014, the EPA announced regulations for fossil-fuel-powered electricity generating units. The required emissions reductions vary by state but will lead to an average reduction of about 30 percent from these plants. Owing to the high carbon intensity of coal, these new regulations will make it quite difficult to operate conventional coal-fired power plants. These regulations will be challenged in court, but, if enacted, they represent the first legitimate effort by the United States to reduce emissions.

Around the same time, the IPCC released its Fifth Assessment Report. It concluded that, "It is *extremely likely* that human influence has been the dominant cause of the observed warming since the mid-20th century." This continues the trend towards strengthening the attribution statement – it now uses the words *extremely likely*, which denotes a 95 percent chance (compared to *very likely* (90%) in the Fourth Assessment and *likely* (66%) in the Third Assessment).

## 13.9 The breakthrough: U.S.-China bilateral agreement

In November 2014, a blockbuster climate deal was announced. The United States and China, the two biggest emitters, together responsible for about one-third of emissions, agreed to limit their emissions. The United States agreed to emit 26 to 28 percent less carbon dioxide in 2025 than it did in 2005. This was perhaps not such an amazing commitment: in response to the Copenhagen Accord, the United States had already committed to reducing emissions by 2020 – but this agreement doubled the pace of emissions reductions.

More impressively, China agreed that its emissions would peak before 2030. This was a true shift in policy, as China had previously only talked about greenhouse gas intensity reductions, not emissions reductions. In order to do this, they committed to produce 20 percent of their energy in 2030 from renewable power sources. This requires them to build about 1 GW of renewable power every week for the next fifteen years.

While the deal only covers the United States and China, I believe that future generations will look upon this as a turning point in negotiations over international climate policy. Including the E.U., which already has aggressive emissions reductions targets in place, this deal means more than half of the world economy, which emits more than half of the carbon dioxide, has agreed to significant deflection of their emissions trajectories.

Most importantly, China is the de facto leader of the developing world, and China's agreement to limit emissions in the near term will put enormous pressure on other developing countries to do the same. It will put even more pressure on industrialized countries that refuse to reduce emissions. Over the years, the excuse that China had steadfastly refused to reduce emissions had become one of the most important arguments for those countries – and now that excuse is gone. Countries like Canada and Australia will find it harder now to do nothing about climate change.

## 13.10 Chapter summary

- Scientists have been studying climate change for nearly 200 years, and in that time a successful theory of climate has emerged. This theory is described in Chapters 1–7 of this book.
- The first prediction of human-induced climate change was made by Svante Arrhenius, who recognized in the late nineteenth century that human combustion of fossil fuels might warm the climate. In the late 1930s, Guy Stewart Callendar made the first claim that human-induced global warming had arrived.
- In the 1950s, people realized that humans possessed the power to greatly modify our environment – and not to our benefit. And the economic growth and increases in wealth occurring at that time meant the environment had more value to people, and people had more money to spend to enjoy it.
- In the 1970s and 1980s, the debates over ozone depletion and acid rain were a preview for the debate over climate. Those opposed to action on these problems refined the strategy of the tobacco companies: Cast doubt on the science.
- The first climate treaty was the Framework Convention on Climate Change or FCCC. This treaty enshrined three important principles: 1) "common but differentiated responsibilities," 2) the precautionary principle, and 3) an agreement that the world should limit greenhouse-gas emissions in order to prevent "dangerous" climate change.
- The 1997 Kyoto Protocol included binding reductions of emissions for industrialized countries – these countries had to reduce emissions from 2008 through 2012 by roughly 5 percent below 1990 emissions. There were no restrictions placed on developing countries.
- The 2009 Copenhagen Accord included the agreement that the world's goal should be to avoid 2°C of warming above preindustrial temperatures. In addition, industrialized countries agreed to set their own emissions reduction targets for the year 2020.
- In late 2014, the United States and China mutually agreed to limit their emissions. This may be viewed in the future as a turning point in the efforts to get an international agreement to limit emissions.

# Terms

Chlorofluorocarbons, or CFCs
Climate skeptics
Common but differentiated responsibilities
Copenhagen Accord
Equity
Framework Convention on Climate Change, or FCCC
Killer smog of London
Kyoto Protocol
Montreal Protocol
Ozone hole
Precautionary principle
Tobacco strategy

# Additional reading

S. R. Weart, *The Discovery of Global Warming*, 2nd ed. (Cambridge, MA: Harvard University Press, 2008). This is an accessible, well-written historical timeline of primary developments in the science of climate change, from the 1800s through the formation of the modern consensus about the predominantly human cause of recent climate change as expressed in the recent IPCC reports (accessible online at www.aip.org/history/climate/index.htm).

N. Oreskes and E. M. Conway, *Merchants of Doubt: How a Handful of Scientists Obscured the Truth on Issues from Tobacco Smoke to Global Warming* (London: Bloomsbury Press, 2010). As I stated in Chapter 1, this important book explains how deception is used to misguide the public on various matters, from the risks of smoking to ozone depletion to the reality of global warming.

See www.andrewdessler.com/chapter13 for additional resources for this chapter.

# Problems

1. Your roommates have a party when you are out of town.
   a) When you return, the apartment is a mess and they ask you to help clean it up. Do you help them?
   b) Under what conditions might you offer to help? What might they offer you to get you to help them?
   c) How is this situation analogous to the debate between developing and industrialized countries over mitigation efforts?
2. Who was the first person to discover that climate could change? Who was the first person to predict that human emissions of carbon dioxide might warm the climate? Who first claimed that human-induced climate change was occurring?

3. Do an Internet search and find some Web sites skeptical of mainstream climate science. List three of the claims they make about the science of climate change. Given what we have covered in the first twelve chapters of this book, are these arguments convincing?

4. What are the four important components of the Framework Convention on Climate Change?

5. What difference is there in how we view the environment today versus how people who lived in the nineteenth century viewed it? What are the factors that caused the change?

6. Explain the precautionary principle. Can you think of an example in your life when you have applied the concept (or explicitly not applied it)?

7. a) Explain the concept of equity as it was described in this chapter.
   b) How was it implemented in the Montreal Protocol?
   c) Give an example of how it might be implemented in a climate agreement.

8. Under the Copenhagen Accord, China agreed to reduce its greenhouse gas intensity by 45 percent in 2020 (relative to a base year of 2005).
   a) What annual growth rate gives you a 45 percent decrease over fifteen years?
   b) If affluence grows at 7 percent/year and population grows at 1 percent/year, what would be the expected change in total emissions over the entire time period?

9. How does the bilateral U.S.-China climate agreement incorporate the concept of equity?

# 14

# Putting it together: A long-term policy to address climate change

We have now reached the final chapter on our trip through the problem of modern climate change. In the previous thirteen chapters, we explored the fundamental physics that leads us to confidently conclude that humans are now changing the climate and that continuing to add greenhouse gases to the atmosphere could bring serious changes to our climate over the next century and beyond. We are not certain how bad this climate change will be, but the upper end of the range (global and annual average warming of 4 °C or more over the twenty-first century) includes warming large enough for the experts to consider its impacts to be potentially catastrophic. Even the lower end of the range, about 1 °C, is more warming than occurred during the twentieth century and will be challenging for the world's poorest as well as our most vulnerable ecosystems. We have also explored a number of possible responses to this risk, including mitigation, adaptation, and geoengineering. We have even touched briefly on the political debate over climate change.

In this chapter, I will discuss the elements of an effective response to climate change. I will also show the step-by-step logic that underlies the most commonly suggested policies for addressing climate change. Our choice of climate and energy policy must reflect the science as well as the economic trade-offs and moral judgments about the alternatives.

## 14.1 What makes climate change such a difficult problem?

I hope that, by this point, you recognize that the physics of the climate problem is actually pretty simple and we can have high confidence that humans are altering the climate. While there is uncertainty in the science, it is mainly about how bad the impacts of climate change will be, not whether humans are altering the climate or whether negative impacts will occur.

Despite the scientific certainty, it has been difficult for the world to get together and address this threat. To see what makes this such a hard problem, let us first consider a different, well-known problem: terrorism. Probably the most famous terrorist attack in history took place on September 11, 2001, when terrorists hijacked and crashed four U.S. airliners, flying three of them into buildings and killing about 3,000 people. In that attack, there was a clear perpetrator (the hijackers and their facilitators) and a clear victim (the people killed and survivors who are terrorized); the cause and effect were clear (the hijackers took control of the planes and crashed them, killing the victims and terrorizing survivors); the effect was immediate (the impacts occurred immediately after the hijackings); and there was a clear intent to do something bad (the hijackers intended to harm people and society).

The response to this attack was immediate and strong. The United States invaded Afghanistan and toppled the government, who had aided the 9/11 terrorists. The attack also provided a key motivation for the 2003 U.S. invasion of Iraq. Given the strong and aggressive response to the terrorist attacks of 9/11, why does another existential threat, climate change, get so little traction in U.S. policy discussions? The answer is that climate is a more diffuse problem than terrorism. First, we cannot pick out a single perpetrator. The people responsible are just about everyone who has lived since the industrial revolution, including you and me. In addition, no single person emits enough greenhouse gases to be an issue. If you were the only emitter, there would be no problem. It is only because there are billions of people on the Earth that emissions are causing the climate to change.

Second, the victims of climate change are also dispersed in time. While we are already altering the climate, the most severe impacts of climate change will be felt toward the end of the twenty-first century and beyond, when the climate is several degrees Celsius warmer than today. This means that many of the people who will be harmed by climate change have not even been born. There is also a disconnect between cause and effect in space: many of the impacted will live in very poor countries, while most emissions come from richer countries. The upshot of this is that many people alive today do not feel personally threatened by climate change the way they do by terrorism. This is probably the most important reason that climate is not a higher policy priority than it is.

Third, the link between cause and effect is not as straightforward as it was for the 9/11 attacks. While we can reasonably attribute rises in global average temperatures to human activities, it is more difficult to attribute individual events that really affect people, such as individual heat waves, droughts, floods, and severe storms to climate change due to human activities.

Finally, those who cause climate change (i.e., everyone) do not intend to cause it. You do not get into your car with the intent of causing climate change, and you would still drive your car even if it did not produce greenhouse gas emissions (in fact, most people would prefer it if their car did not cause climate change). Rather, emissions and the resulting climate change are a side effect of economic activities that are otherwise virtuous – this means that restrictions on emissions carry the possibility of harming our economy, so one must balance climate policies against restrictions in economic activity. Overall, these differences make the climate problem harder to deal with than more immediate threats.

## 14.2 Decisions under uncertainty: Should we reduce emissions?

Despite the difficult philosophical aspects of the problem discussed in the last section, we nevertheless face a choice with climate change: to act or not to act. That decision will be based on answers to a few key questions. Some of these are scientific (How much warming will we experience? What will the impacts be?), some are economic (How expensive will it be to respond?), and some are (at least partially) moral (How bad will the impacts be? Is geoengineering an acceptable response?). Ultimately, we don't have precise answers to any of these questions. Such uncertainty is not unusual in important policy decisions. Most important decisions (Should we invade Iraq? Should we cut taxes? Should we implement universal health care?) are made with incomplete information. So how can we make a decision in the face of this uncertainty? One way to think about this problem is to consider the following two arguments:

- Because the worst-case scenario of climate change is so serious, we must take action now to reduce emissions, even though we don't know exactly how bad climate change will be.
- Because of the high cost of reducing emissions, we must be certain that climate change is serious before we take action.

Both statements argue that we must err on the side of caution in order to avoid a bad outcome. However, the bad outcome is different in these two arguments. In the first argument, the bad outcome is severe damage from climate change, whereas in the second it is severe economic damage from responding to climate change.

So which of these arguments is correct? We can gain some insight into how to think about this by looking at some familiar examples of decisions in the face of uncertainty. First, consider a criminal trial. To convict someone of a crime, a jury must be convinced that the defendant is guilty beyond a reasonable doubt. The reason for this standard is that we, as a society, have decided that it is better to acquit a guilty person than convict an innocent one.

Put another way, there are two errors a jury can make. They can convict an innocent person or they can acquit a guilty one. These errors are not equally bad – we judge that it is worse to convict an innocent person. So the standard of conviction ("guilty beyond a reasonable doubt") is set to minimize the possibility of making the worse mistake. In doing so, we increase the chance that we make the other mistake, acquitting a guilty person.

Another example occurs in deliberations concerning national defense. For example, former Vice President Dick Cheney famously said, "If there's a one-percent chance that Pakistani scientists are helping al Qaeda build or develop a nuclear weapon, we have to treat it as a certainty in terms of our response."[1] As in our jury example, there are two errors here that we could make. We could respond as if al Qaeda had a nuclear weapon, but it turns out they do not have a nuclear weapon. Or we do not respond, and it turns out they do have one. In most deliberations about national defense, being unprepared for a threat is judged to be a worse error than to respond to a threat that never materializes. That is the fundamental judgment that Cheney is making here.

So how do we think about climate change? Must we be certain beyond a reasonable doubt that climate change is a serious threat to mankind before taking action to

---

[1] Quoted in R. Susskind, *The One Percent Doctrine: Deep Inside America's Pursuit of Its Enemies since 9/11* (New York: Simon & Schuster, 2006).

reduce emissions? Or should we take action to reduce emissions even if there is just a 1 percent chance that it is a serious threat? This question boils down to your judgment of which error is worse: reducing emissions unnecessarily because climate change turns out to be a minor threat or not reducing emissions and climate change turns out to be a serious threat.

Suppose that climate change turns out to be a minor problem. In that case, an aggressive mitigation program would impose costs as we rebuild our energy infrastructure from fossil fuels to renewables and other climate-safe energy sources. How bad would this be? Switching from fossil fuels to climate-safe energy has advantages completely unrelated to climate, such as reductions in air pollution. Moreover, because costs of transitioning to climate-safe energy would be spread over the next several decades, at least some of the cost can be avoided by scaling back future efforts once we learn they are unnecessary. Furthermore, fossil fuels will be exhausted in the next century or so. Thus, switching away from fossil fuels is inevitable and these costs are going to be paid eventually. The bottom line is that it is hard to imagine a person living in Year 2100 being upset that we switched from fossil fuels to climate-safe energy sources.

Now suppose that climate change turns out to follow the worst-case scenario. If we do nothing to reduce emissions, we doom the planet to much warmer temperatures for the next millennium and beyond. This could impose catastrophic costs on our society (both economic and moral). In this case, it is easy to imagine that a person living in Year 2100 would be furious that we did nothing to address a problem that we clearly saw coming. In the end, it is difficult to make the argument that taking too much action on climate change is a worse error than taking too little. This would suggest a standard closer to Cheney's, that the risk of climate change justifies action to reduce emissions, even in the face of significant uncertainty.

Another factor that enters into decisions under uncertainty is irreversibility. If an action you take is irreversible, you have to be more certain that it is the right action than for a decision that is easily reversible. That is why, for example, inmates on death row in the United States are allowed many appeals to their death sentence – executing someone is as irreversible an action as there is, so you want to be as sure as possible that you have executing the right person. Just putting someone in jail, in contrast, is reversible – if you realize later that you have made a mistake, you can simply release that person.

Reducing emissions is a reversible decision. If we decide later that climate change is not that serious, we can always change our policies and increase our consumption of fossil fuels. And the investments we make in alternative energy sources, such as solar and wind, are reversible over a few decades.

But the converse is not true. Once you emit carbon dioxide, there is no practical way to remove it from the atmosphere. Instead, you have committed the planet to millennia of higher temperatures or geoengineering. Many of the impacts of climate change, such as extinction of species, sea-level rise, and loss of the Greenland or Antarctic ice sheets, are irreversible on time scales that we care about. Thus, the irreversibility of emitted carbon dioxide and its associated climate change tends to favor taking action to reduce emissions. In the end, most people who have seriously looked at the problem, including almost every world government, have concluded that action on climate change, in particular the reduction in emissions, is justified given the risks.

# 14.3 Picking a long-term goal

Any emissions-reduction effort requires a long-term goal. The deeper the cuts in emissions, the less climate change we will eventually experience – but the more expensive those cuts will be. This is the trade-off, and we want to pick a target that avoids the worst climate change but at a cost that is manageable and does not interfere with other policy goals, such as poverty reduction.

The Framework Convention on Climate Change says that we should strive to avoid "dangerous" climate change. But what is *dangerous?* It is not a scientific term, so science provides input, but cannot settle the issue. Rather, it is a value judgment that takes science into account as well as our views on topics such as risk, poverty, environmental stewardship, government regulation, and many other contentious topics. There are various ways to determine a long-term goal, and I discuss two of them in this section.

## 14.3.1 Cost versus benefits

In Chapter 12, we described how a company would respond if a price were put on emissions through a carbon tax or cap-and-trade system. As described there, companies compare costs and benefits to choose the emissions reduction that maximizes their net benefit (benefits minus cost). This occurs when the cost of reducing one more unit (the marginal cost) equals the cost of emitting that unit (either through a tax on that unit or the cost of a permit to emit that unit).

Societies can make emissions reductions decisions in an analogous way: by comparing the cost of various levels of emissions reduction to the benefits obtained by making those reductions. Imagine, for example, that our economy can reduce emissions of carbon by 1 ton for $5, but we get $50 of benefits by avoiding the climate impacts of that ton. From a societal standpoint, that is a no-brainer – we should certainly not emit that ton. Now imagine that the next ton of emissions can be avoided for $6, and avoiding emission of this ton delivers $49 of benefits. Again, it is clear that we should pay to not emit that ton either. As we cut emissions deeper, the cost of eliminating each subsequent ton (the marginal cost) rises, while the benefits from avoiding that ton (the marginal benefit) decline. This is the law of diminishing returns, which we explored in Chapter 12.

The reduction that gives us the largest net benefit would be our preferred goal, and this occurs when emissions are reduced until the cost of reducing one more ton equals the benefit from avoiding that ton. Even though this may be conceptually straightforward, the actual calculation is not. For example, we know how much renewable energy costs today, so we can estimate the cost of replacing fossil fuel energy. But we also know that putting a price on carbon will spur development of new technologies by providing a financial incentive that does not exist in today's economy. That this will happen is not in doubt – the question is how fast the innovation will take place. If we assume that innovation responds rapidly to putting a price on emissions, then the cost of reducing emissions will be much lower than if we do not make that assumption. Depending on what assumptions are made for this and other uncertainties, estimates of the costs of reducing emissions cover a wide range; some analyses conclude that it will be quite cheap whereas others conclude that it will be ruinously expensive.

Estimating the benefits from avoiding climate change is even more difficult. First, we do not at present have the ability to predict changes in temperature and precipitation at the regional scales required for detailed estimates of impacts. Second, converting estimated changes in climate into a dollar figure can be difficult and arbitrary. For goods and services that are traded in markets (e.g., food, lumber, recreational skiing), calculating the economic loss due to climate change is relatively straightforward. For things that are not traded in markets, however, estimates of the cost of climate change are far more arbitrary. Take, for example, the extinction of polar bears. Polar bears do not contribute much to the global economy, so their extinction would likely have a negligible financial cost. But many people nevertheless value polar bears and would view their loss as a significant harm. Economic analyses can attempt to quantify the value of polar bears by using methodology that is neither uniform nor entirely satisfactory, or analyses can ignore their loss, which implicitly assigns it a value of zero – which is also unsatisfactory. Given the many problems in estimating the cost of climate change, it should come as no surprise that there is also a wide range of cost estimates.

Another problem in estimating the costs of climate change comes from the timing of climate impacts. If 1 ton of carbon is emitted into the atmosphere today, it will warm the planet for centuries to come and will cause impacts over that entire time. But the cost of avoiding those impacts – by not emitting that ton – must be paid today. To compare costs that are occurring at different times, we therefore need to convert the value of the climate impacts over the next few hundred years to its value to us today (i.e., its present value). This involves discounting, which we explored in detail in Section 10.4.

Although discounting is conceptually straightforward, the big uncertainty is what discount rate to use. Most analyses in the climate change policy debate use discount rates between 0 percent and 4 percent. The larger the discount rate, the lower the present value of future costs – and we will consequently be willing to pay less to avoid those impacts.

For example, with a discount rate of 0 percent, $1 trillion of climate change damage in 100 years has a present value of $1 trillion. We should therefore be willing to pay up to $1 trillion dollars today to avoid it. In contrast, if we select a discount rate of 4 percent, then $1 trillion of climate change damage in 100 years has a present value of $20 billion – meaning we would only be willing to pay $20 billion to avoid those impacts.

This is even more problematic for impacts occurring far in the future. As we explored in Chapter 8, our emissions this century will commit the planet to warming and climate impacts over the coming millennium and beyond. The present value of $1 trillion of climate impacts in 1,000 years at a 0 percent discount rate is $1 trillion;

at 4 percent, it is about 0.001 cent. The staggering difference between these present values has significant implications for how much emissions reduction is optimal.

Together, uncertainties in estimates of the costs of reducing emissions, costs of climate impacts, and the discount rate lead to a wide range of estimates of how deeply to cut emissions in order to maximize net benefits. Some estimates are for deep, immediate cuts, while others call for more gradual ramping down of emissions over decades. Because of this, economics is limited in its ability to prescribe a quantitative response to climate change.

Nonetheless, there are some things that all economic analyses agree on, and on those points we can have high confidence. There is widespread agreement that reductions in emissions make sense and that carbon pricing should be the centerpiece of the policy. In addition, there is agreement that the price on emissions should rise with time, eventually becoming very high. The net result is that emissions can be allowed to grow in the near term (for perhaps a decade or so) before emissions must actually begin to decline. The exact timing of the emissions peak is determined by how far we decide emissions must be cut – the more climate change we want to avoid, the nearer in time the cuts must begin. It is also important that we have near-universal participation by all countries in any climate regime, in order to make the necessary reductions at the lowest cost.

Economic analysis suffers from another problem: It looks at aggregate costs and benefits but not their distribution. It does not take into account that many of the hardest-hit regions are also the poorest regions of the world – regions that have contributed little to climate change and have the least resources available to address climate. Many would view that outcome as fundamentally unfair, a factor not considered in economics. So although economics tells us what the most efficient solution is, it tells us nothing about whether the solution is fair or just.

A final problem with economic analysis comes from the problem of catastrophe, including worst-case outcomes such as abrupt climate changes, mass starvation, or even human extinction. Economic analyses struggle to assign a monetary value to a small and hard-to-quantify risk of truly terrible outcomes like these. As a result, most economic analyses ignore the worst-case scenario and focus on the most likely outcomes. This is a significant shortcoming: For many people, the mere possibility of true catastrophe in the next few centuries provides sufficient motivation for us to address climate change now.

## 14.3.2 Target: 2 °C

Given all of the problems with economic analyses, expecting them to quantitatively determine the optimal amount of mitigation is not realistic. A simpler way is to choose some threshold in the climate system that we should avoid. The threshold should give us a good chance to avoid serious climate impacts, but it should be relaxed enough that it is politically and economically acceptable. Over the past few years, a consensus has grown around limiting warming to 2 °C above preindustrial temperatures. This is what was adopted, for example, at the FCCC's Copenhagen meeting in 2009. The scientific argument in favor of this threshold is that modern human society, with megacities and large-scale industrial food production, developed during a period of relatively small climate variations – less than 2 °C. This stable climate provided the conditions (e.g., good weather for agriculture, robust freshwater supplies) under which human society flourished.

Ultimately, though, this 2 °C is arbitrary. There is no scientific analysis that proves that 2 °C is the most appropriate target, nor any reason to think that warming slightly below this threshold is much better than warming slightly above. And we could just as easily have chosen a completely different metric, such as a limit in rate of warming (how many degrees per century we were willing to accept) or a limit in atmospheric carbon dioxide. Some advocates, for example, argue that we should limit atmospheric carbon dioxide to 350 or 450 ppm. However, as of 2014, consensus has arisen around the 2 °C target, and it has most of the momentum in policy discussions.

## 14.4 How do we get there?

### 14.4.1 The physics of a 2 °C limit

This is a challenging target – we have already experienced approximately 0.8 °C of warming above the preindustrial level, and we are committed to another few tenths of a degree Celsius of warming from emissions that have already occurred. This leaves us a bit less than 1 °C of warming that it is still physically possible to avoid.

Avoiding all of that 1 °C is impossible because that would require turning off most fossil-fuel energy immediately. That would be too expensive – the best we can hope for is to phase fossil fuel power out over the next few decades, during which time atmospheric carbon dioxide will continue to build up. But how fast do we need to turn off fossil fuels?

A useful way to think about this problem is in terms of a carbon budget. Because carbon dioxide remains in the atmosphere for so long (discussed in Chapter 5), the peak warming we will eventually experience is proportional to cumulative emissions of carbon dioxide. Climate models suggest that the climate will warm about 2 °C for every trillion tons of carbon emitted (note: 1 trillion tons of carbon = 1,000 billion tons = 1,000 GtC). Humans have already emitted about 500 GtC since the industrial revolution, so we can stay under 2 °C warming if we limit future emissions to 500 GtC. This is a tight budget: While it took us 250 years to plow through the first half of our carbon budget, given the exponential growth of the rate of emissions, we are on schedule to plow through the second half of our carbon budget in the next few decades – unless we take action to reduce emissions.

Other greenhouse gases matter, of course, but carbon dioxide is the most important because it remains in the atmosphere for so long (discussed in Chapter 5). For a gas like methane, with a ten-year lifetime, once you reduce emissions, the atmospheric perturbation due to humans, and the associated radiative forcing and warming, will fall to zero within a few decades.

As with all budget problems like this, we have three options: we can start cutting emissions now and reduce our emissions (relatively) gradually, or we can continue to do nothing, let emissions grow each year, and then make drastic emissions cuts later.

Or we can do nothing now and do nothing later and simply accept large changes in our climate.

A few numbers will help clarify our choices. If we choose to do nothing, warming over the twenty-first century could be 4°C (this is the prediction of the RCP8.5 scenario, see Figure 8.5), corresponding to about 5°C warming above preindustrial. That is about as much warming as we have had since the end of the last ice age. And warming is expected to continue well beyond 2100 under this scenario. Given such significant warming, the risks of severe climate impacts would be considerable.

If we decide that we want to keep global warming below 2°C – and we decide we want to start immediately – what level of effort would be required? In this case, we can keep cumulative emissions below 500 GtC if we reduce emissions by 1.5 percent/year. So how do we do that? I noted in Chapter 8 how emissions can be thought of as the product of population, affluence, and technology (Equation 8.1: $I = P \times A \times T$; if you do not remember this, now is a good time to review Section 8.1). We expect the world's population to continue to increase in the future, and we also expect that people will become richer. More quantitatively, we can expect the product of population times affluence ($P \times A$) to increase at 2 to 4 percent/year. This means that the technology term (i.e., the greenhouse gas intensity) must decline by 3.5 to 5.5 percent/year in order for overall emissions to decline by 1.5 percent/year.

Again thinking back to Chapter 8, greenhouse gas intensity is equal to the energy intensity times the carbon intensity (equation 8.2). We expect that energy intensity, a measure of how efficiently an economy uses energy, to continue to decrease by about 1 percent/year. Thus, the bulk of our emissions reduction comes from the carbon intensity term, which must decrease at 2.5 to 4.5 percent/year.

The carbon intensity term is a measure of the technology used to generate electricity. To have it decline by 2.5 to 4.5 percent/year will require construction of about 1 gigawatt (1 GW = $10^{12}$ W) of carbon-free energy every day for the next century. For scale, 1 GW is about the power produced by a coal-fired power plant, a large hydroelectric dam, a few hundred large wind turbines, or a square of solar panels that is a few kilometers on each side.

It is easy to conclude that this task is too big and that we cannot possibly build this much carbon-free energy. But the power of human industrial society is also immense – it is, after all, big enough to shift our climate. As an example, consider the following fact: human population increased by about 1 billion in the last thirteen years (corresponding to a population increase of 200,000 people/day) and most of the population increase occurred in urban areas. This means that humans have been building the equivalent of a city of one million people every five days for the last decade. And we will continue this rate of urban construction for the next several decades, at least. Given that, switching to carbon-free energy does not seem so daunting, and it is clear that we *could* build enough carbon-free energy if we chose to do so.

If we wait to start reducing emissions until, say, 2030, we have to reduce greenhouse gas intensity faster – about 6 to 11 percent/year – to stay within our 2°C carbon budget. If we wait until 2040 to start reducing, then we have to reduce the greenhouse gas intensity even faster – by 13 to 50 percent/year. These all correspond to much higher rates of building carbon-free energy sources and higher overall costs. Given how difficult it has been for the world to agree to do anything at all about climate, I am skeptical that the world would ever agree to such high rates of emissions reductions. In such a situation, it seems likely to me that we would turn to geoengineering to buy time to let a more modest mitigation program take effect.

## 14.4.2 How to get there from here

As described in the previous section, we know what needs to happen from a scientific viewpoint in order to stabilize the climate at 2 °C above the preindustrial level – we need to cut emissions of greenhouse gases, particularly carbon dioxide, and we need to start as soon as possible. It is also clear that all countries need to participate in this endeavour. If one or a few big countries choose not to participate, then participating countries will have to make larger and faster cuts in emissions. This is more expensive for the overall economy, and it is particularly expensive to those making the steep reductions. This leads to more economic pain for those participants, potentially eroding political support in those countries.

But getting all nations to agree to reduce greenhouse gas emissions is difficult because, in the international arena, no one is in charge. There is no world government with the authority to implement and enforce policy or to compel governments to participate. Rather, international policy is made by negotiation among national representatives. This process is weaker, more cumbersome, and slower than national policy-making, but it is all that is available to respond to global problems such as climate change.

The goal of international climate negotiations is for each country to commit to some level of effort to address climate change. These national commitments could be performance targets, such as limits on national emissions, leaving it up to the individual governments how to achieve them. Alternatively, governments could commit to enact national policies, such as emission taxes or other forms of regulation. Or they could submit to international processes to motivate and enable national actions, such as reporting and exchange of information, assessment of national policies, or reviews of their progress.

Thus far, national emissions targets, in which countries agree to reduce emissions by some specified amount in some specified time period, have been the most common form of international mitigation commitment, used in both the Framework Convention on Climate Change and the Kyoto Protocol, as well as several other major environmental treaties. National targets have the advantages of being simple, clear, and familiar. Moreover, by defining clear responsibilities but leaving the means of implementation up to national governments, targets let governments be held accountable for their commitments with minimal intrusion on their sovereignty.

But national targets also have serious disadvantages. Because most emissions come not from governments but from citizens and businesses, a national target has no concrete effect until implemented in domestic policy. But the outcomes of domestic policies are uncertain at the outset, so governments cannot know in advance how difficult or costly it will be to meet an emissions target, or even whether a particular target is achievable. This uncertainty about targets' attainability introduces a sharp tension into international target negotiations. If governments face only minor consequences for missing a target, they have little incentive for serious efforts to meet it. But if the consequences are severe, governments will likely only agree to targets they are highly confident of meeting – weak targets, or ones with wide loopholes. Moreover, while clear, demanding targets can be good motivators, they do this best when nations are near the boundary between meeting and not meeting them. Incentives to exceed a target you already expect to meet, or to narrow the gap when you are clearly falling short, are much weaker.

But the biggest problem with national targets is assigning burdens and costs among nations. The most intense disputes in this category are between developing countries and the rich, industrialized countries. Developing countries argue that the burden for near-term cuts should fall on the world's rich countries, and they advance several reasons in support of this. The first is historical responsibility. Most of the increase in carbon dioxide in the atmosphere over the past 250 years is due to emissions from the industrialized countries. In fact, the world's richest countries are rich because they have consumed a lot of energy. Thus, it makes sense for them to have a greater responsibility for taking the first steps toward addressing the problem.

Second, the industrialized countries are wealthier and have more resources available to make the investments required to reduce emissions. By having the rich countries reduce emissions first, development of new technologies would be expected to lower the cost for the developing countries of their eventual emissions reductions.

Third, climate change policies cannot work against poverty reduction. Two billion people today live in intense poverty. Lifting them out of that state requires energy, so anything that makes consuming energy harder or more expensive for the poorest will also make it harder to reduce poverty. This is a strong moral argument that the poorest countries in the world should not be asked to reduce emissions. As they become richer, they can contribute to emissions reductions.

The argument that the rich world should move first has received mixed reviews in the rich world. In 1997, in response to the negotiation of the Kyoto Protocol, the U.S. Senate passed a resolution with a vote of 95–0 stating that the United States should not mandate new commitments to limit or reduce greenhouse gas emissions unless the agreement also includes commitments to limit or reduce greenhouse gas emissions by developing countries.

Over the following seventeen years, this issue was never really settled. Developing countries refused to agree to reduce emissions, although they agreed that they would have to do so eventually. The European Union moved ahead on its own aggressive emissions reduction targets, while Canada and Australia essentially abandoned any formal targets. The United States was somewhere in between. During Obama's second term, various executive orders (which do not require congressional approval) were enacted that are expected to reduce emissions, although they fell short of the kind of comprehensive energy policy that would significantly reduce U.S. emissions. Nevertheless, U.S. emissions have been declining since the mid-2000s. This is mainly due to three factors: 1) the great recession, which slowed the growth of affluence, 2) a decrease in energy intensity due to the long-term increases in energy efficiency of the economy, and 3) the decrease in carbon intensity due to a shift away from coal following the rise of new drilling technology (i.e, fracking), which flooded the energy market with cheap shale gas.

As I write this in late 2014, important events have signalled a possible breakthrough in international negotiations. As discussed in the previous chapter, the United States and China recently announced a bilateral agreement to rein in emissions: The United States intends to reduce its emissions by 26 to 28 percent below its 2005 level by 2025. China intends to achieve the peaking of carbon dioxide emissions around 2030 and to make efforts to peak early. China also intends to increase the share of non-fossil fuels in primary energy consumption to around 20 percent by 2030, which will require the construction of about 1 GW of renewable power every week between now and then. The agreement between the de facto leaders of the industrialized and developing worlds to limit emissions has the possibility of breaking the rich versus poor divide that has stymied negotiations over the past twenty years.

### 14.4.3 What policies should look like

The national targets that result from international negotiations do nothing until implemented in domestic policy. So what would a domestic policy look like? The first and most important action every nation can take is to put a price on emissions of carbon dioxide and other greenhouse gases (why this is necessary was covered in Chapter 11). Economist William Nordhaus put it this way:

> Whether someone is serious about tackling the global warming problem can be readily gauged by listening to what he or she says about the carbon price. Suppose you hear a public figure who speaks eloquently of the perils of global warming and proposes that the nation should move urgently to slow climate change. Suppose that person proposes regulating the fuel efficiency of cars, or requiring high-efficiency light bulbs, or subsidizing ethanol, or providing research support for solar power – but nowhere does the proposal raise the price of carbon. You should conclude that the proposal is not really serious and does not recognize the central economic message about how to slow climate change. To a first approximation, raising the price of carbon is a necessary and sufficient step for tackling global warming. The rest is at best rhetoric and may actually be harmful in inducing economic inefficiencies.[2]

Second, although a price on carbon is crucial, there are some economic sectors where great progress can be made rapidly but where expected progress with just a price on carbon will be slow. For these sectors, efficiency standards and other incentives can be implemented to encourage careful energy use. One example is fuel mileage standards for automobiles. In addition, restrictions on activities that are particularly unfriendly to the climate, such as the burning of coal, might also be considered.

Third, countries should fund the research and development of new technologies. Although it may be possible to solve the climate problem with existing technology, it is also clear that new and improved technologies can ease the transition as well as reduce the cost. An example of transformative technology would be a breakthrough in battery energy storage. This would reduce the impact of the intermittency of renewable energy – e.g., solar energy generated during the day could be stored and then fed into the grid at night. Another example would be an "artificial leaf" that

---

[2] See Nordhaus (2008), p. 22.

takes sunlight and uses it to split water into hydrogen and oxygen; the hydrogen could then be burned to produce energy.

Fourth, prepare to adapt to climate change. Regardless of what actions we take now, the globe will continue to warm for decades. And to the extent that this warming cannot be avoided, we must adapt to it. Thus, adaptation must necessarily be a part of our response to climate change. Anticipatory adaptation is cheapest, so we should begin to incorporate the reality of climate change into any plans for the future. This is particularly true for investments in long-lived infrastructure. When we build a road, airport, power plant, or the like, we must make sure that it is resilient to any reasonable changes in the climate. In addition, not every country has the resources to adapt, so mechanisms of providing international aid may have to be implemented.

Fifth, there is also the unfortunate possibility that the world will not get its act together to reduce emissions. If the world does nothing to reduce emissions in the next few decades, then geoengineering may be our only way to avoid truly disastrous warming. We should prepare today for this by researching the geoengineering options in order to determine which are most likely to work and what the negative side effects might be.

Sixth, we must also realize that whatever policy we adopt now will probably not be exactly the right long-term policy. Because of this, both international agreements and the domestic policies that flow from them must be reviewed and amended as new information about the science of climate change and new technological developments arise.

In the end, we do not confidently know how hard it will be to reduce our emissions and make the transition to a fossil-fuel-free future. In this way, the climate change challenge is not unique; we almost never know in advance how much it will cost to comply with environmental regulations. Before regulations to reduce ozone depletion were passed in the 1980s, for example, some advocates predicted that the regulations would cause an economic apocalypse, with people in the developed world having to get rid of their air conditioners and millions in the developing world dying because of a lack of food refrigeration.

It turned out that innovation in response to the threat of regulation led to the development of substitutes for the ozone-depleting chemicals so cheap that, when the new chemicals replaced the older ones, virtually no one noticed. The hope with climate change is that, once a price is put on carbon emissions, innovation by the private sector will produce breakthrough technologies that allow us to make reductions in emissions cheaply and with minimal economic disruption. Whether or not this will happen is impossible to know until we try.

## 14.4.3 What can you do?

If you have decided that climate change is something that we need to address, you may be wondering what you as an individual can do. There certainly are personal choices you can make that will reduce your share of emissions of greenhouse gases. Some of these choices will not only reduce emissions but also save you money and benefit you in other ways (e.g., take public transportation or walk instead of driving, eat less meat). Other choices may require upfront costs but will pay for themselves over subsequent years (e.g., add insulation to your attic or switch to LED lighting). Some choices are difficult to justify on economic grounds alone (e.g., install photovoltaic solar panels on your house[3]).

You should certainly undertake all of the voluntary individual actions that you can. But these individual actions are not going to lead to the emissions reductions necessary to stabilize the climate. Those will require collective, coordinated action at both the national and international levels. That is why the single most important thing you can do is become politically active – write letters to your representatives, participate in rallies, talk to your friends and neighbors, and vote for politicians who support action on climate.

## 14.5 A few final thoughts

In this book, I have tried to give a comprehensive overview of the climate change problem. Unlike what you might hear in the public debate, much of the science of climate change is extremely solid. There is no question that, when you add a greenhouse gas to the atmosphere, the planet will warm (Chapters 4 and 6). There is no question that human activities are increasing the amount of greenhouse gas in our atmosphere (Chapter 5). There is no question that the Earth is currently warming (Chapter 2), and it is warming about as much as you would expect from the addition of greenhouse gases (Chapter 7). This science is not new – much of it is a century or more old and has stood the test of time.

Other aspects of the problem are less certain. Quantitative projections of future climate change at regional scales still contain significant uncertainty; uncertainties also exist in the emissions scenario the world will follow (Chapter 8). Moreover, this uncertainty is magnified by uncertainty in how this warming will impact humans and those aspects of the environment that we care about. However, one conclusion is clear: If climate change falls toward the upper end of the predicted range, we will truly be remaking the face of the planet, and the results may be dire, perhaps even catastrophic (Chapter 9).

We know how to solve this problem (Chapter 11, 12, and 14), but we do not know how hard and expensive it will be – and we will not know until we try. Paralyzed by this uncertainty, the world has made little progress in solving this problem, despite decades of warnings from scientists (Chapter 13).

I do not know what the future holds. But I do know that, if we are going to navigate the coupled problems of energy and climate, we are going to need people like you to get involved in all parts of the problem: the political, the economic, and the scientific. Given the enormous creativity and inventiveness of humans, there is no question that we *can* solve the problem. I encourage you to get involved to ensure that we do.

---

[3]  This might make financial sense if you receive enough of a subsidy from the government.

## Additional reading

N. Stern, *The Economics of Climate Change: The Stern Review* (Cambridge: Cambridge University Press), 2007; W. Nordhaus, *The Climate Casino: Risk, Uncertainty, and Economics for a Warming World* (New Haven, CT: Yale University Press), 2013. These two economic analyses compare the costs and benefits of action on climate change. Both conclude that action is required, but they differ on how much action, mainly as a result of differences in the discount rate. Stern advocates strong action immediately, whereas Nordhaus advocates a slower ramping down on emissions.

P. Krugman, "Building a Green Economy," *New York Times Magazine*, April 7, 2010. This offers a clear and concise summary of the economics of climate change policy and a critique of cost-benefit calculations on climate change (download at www.nytimes.com/2010/04/11/magazine/11Economy-t.html).

A. E. Dessler and E. A. Parson, *The Science and Politics of Global Climate Change: A Guide to the Debate*, 2nd ed. (Cambridge: Cambridge University Press, 2010). Chapter 5 of that book covers the uncertainty question; it also outlines in detail the elements of an effective international response to climate change as well as how we might get there.

S. M. Gardiner, *A Perfect Moral Storm: The Ethical Tragedy of Climate Change* (Oxford: Oxford University Press, 2013). This is an accessible primer describing the ethical issues of the climate change problem.

## Problems

1. What is the single most important thing the world needs to do to address climate change? Why?

2. A friend argues, "We must be certain climate change is a problem before we take action." Another friend argues, "We must take action if the slightest chance exists that climate change could be catastrophic." How do you determine which one is right? Which one (if either) do you judge to be correct?

3. How does irreversibility of the choices affect policy decisions?

4. Imagine that your hometown is always at risk of being destroyed by some natural disaster (tornado, hurricane, earthquake). How much would you pay each year to eliminate the chance that the disaster would occur in that year? What are some ways you could determine this value?

5. a) Explain conceptually the role that discounting plays in determining climate change policy.

   b) If you change the discount rate from 0 percent to 4 percent, how will this change your policy?

6. Juries in criminal trials are given a standard of evidence that must be crossed in order to find a defendant guilty. What is it? What would the standard be if our society decided that the worse error is to acquit a guilty person?

7. For climate impacts happening in fifty years, how much effort should we make to eliminate those? What about impacts happening in 500 years?

8. Economic analyses struggle to assign a monetary value to a small chance of a truly terrible outcome. To see this, imagine that some otherwise unavoidable activity carries with it a 0.1 percent chance of killing you. How much money would you spend to avoid that activity? What if the risk of death were 1 percent or 10 percent?

# References

Balmaseda, M. A., K. E. Trenberth, and E. Kaellen, "Distinctive Climate Signals in Reanalysis of Global Ocean Heat Content," *Geophysical Research Letters,* 40 (2013), 1754–1759, doi: 10.1002/grl.50382.

Ciais, P., C. Sabine, G. Bala, L. Bopp, V. Brovkin, J. Canadell, A. Chhabra, R. DeFries, J. Galloway, M. Heimann, C. Jones, C. Le Quéré, R. B. Myneni, S. Piao, and P. Thornton, "Carbon and Other Biogeochemical Cycles," in T. F. Stocker, D. Qin, G.-K. Plattner, M. Tignor, S. K. Allen, J. Boschung, A. Nauels, Y. Xia, V. Bex, and P. M. Midgley (eds.), *Climate Change 2013: The Physical Science Basis*. Contribution of Working Group I to the Fifth Assessment Report of the Intergovernmental Panel on Climate Change (Cambridge and New York: Cambridge University Press, 2013), pp. 465–570, doi: 10.1017/CBO9781107415324.015.

Collins, M., R. Knutti, J. Arblaster, J.-L. Dufresne, T. Fichefet, P. Friedlingstein, X. Gao, W. J. Gutowski, T. Johns, G. Krinner, M. Shongwe, C. Tebaldi, A. J. Weaver, and M. Wehner, "Long-Term Climate Change: Projections, Commitments and Irreversibility," in T. F. Stocker, D. Qin, G.-K. Plattner, M. Tignor, S. K. Allen, J. Boschung, A. Nauels, Y. Xia, V. Bex, and P. M. Midgley (eds.), *Climate Change 2013: The Physical Science Basis*. Contribution of Working Group I to the Fifth Assessment Report of the Intergovernmental Panel on Climate Change (Cambridge and New York: Cambridge University Press), pp. 1029–1136, doi: 10.1017/CBO9781107415324.024.

Deffeyes, K. S., *Beyond Oil: The View from Hubbert's Peak* (New York: Hill and Wang, 2006).

Denman, K. L., G. Brasseur, A. Chidthaisong, P. Ciais, P. M. Cox, R. E. Dickinson, D. Hauglustaine, C. Heinze, E. Holland, D. Jacob, U. Lohmann, S. Ramachandran, P. L. da Silva Dias, S. C. Wofsy, and X. Zhang, "Couplings between Changes in the Climate System and Biogeochemistry," in S. Solomon, D. Qin, M. Manning, Z. Chen, M. Marquis, K. B. Averyt, M. Tignor, and H. L. Miller (eds.), *Climate Change 2007: The Physical Science Basis*. Contribution of Working Group I to the Fourth Assessment Report of the Intergovernmental Panel on Climate Change (Cambridge: Cambridge University Press, 2007), 499–588.

Dessler, A. E., and E. A. Parson, *The Science and Politics of Global Climate Change: A Guide to the Debate*, 2nd ed. (Cambridge: Cambridge University Press, 2010).

Forster, P., V. Ramaswamy, P. Artaxo, T. Berntsen, R. Betts, D. W. Fahey, J. Haywood, J. Lean, D. C. Lowe, G. Myhre, J. Nganga, R. Prinn, G. Raga, M. Schulz, and R. Van Dorland, "Changes in Atmospheric Constituents and in Radiative Forcing," in S. Solomon, D. Qin, M. Manning, Z. Chen, M. Marquis, K. B. Averyt, M. Tignor, and H. L. Miller (eds.), *Climate Change 2007: The Physical Science Basis*. Contribution of Working Group I to the Fourth Assessment Report of the

Intergovernmental Panel on Climate Change (Cambridge: Cambridge University Press, 2007), 129–234.

Hansen, J., R. Ruedy, M. Sato, and K. Lo, "Global Surface Temperature Change," *Reviews of Geophysics,* 48 (2010) RG4004, doi: 10.1029/2010RG000345.

Houghton, J., "Madrid 1995: Diagnosing Climate Change," *Nature* 455 (2008), 737–738 (accessible online at http://www.nature.com/nature/journal/v455/n7214/full/455737a.html).

IPCC, "Summary for Policymakers," in S. Solomon, D. Qin, M. Manning, Z. Chen, M. Marquis, K. B. Averyt, M. Tignor, and H. L. Miller (eds.), *Climate Change 2007: The Physical Science Basis.* Contribution of Working Group I to the Fourth Assessment Report of the Intergovernmental Panel on Climate Change (Cambridge: Cambridge University Press, 2007a), 1–18.

IPCC, "Summary for Policymakers," in B. Metz, O. R. Davidson, P. R. Bosch, R. Dave, and L. A. Meyer (eds.), *Climate Change 2007: Mitigation.* Contribution of Working Group III to the Fourth Assessment Report of the Intergovernmental Panel on Climate Change (Cambridge: Cambridge University Press, 2007b), 1–24.

IPCC, "Summary for Policymakers," in T. F. Stocker, D. Qin, G.-K. Plattner, M. Tignor, S. K. Allen, J. Boschung, A. Nauels, Y. Xia, V. Bex, and P. M. Midgley (eds.), *Climate Change 2013: The Physical Science Basis.* Contribution of Working Group I to the Fifth Assessment Report of the Intergovernmental Panel on Climate Change (Cambridge and New York: Cambridge University Press, 2013), 3–32.

Jansen, E., J. Overpeck, K. R. Briffa, J.-C. Duplessy, F. Joos, V. Masson-Delmotte, D. Olago, B. Otto-Bliesner, W. R. Peltier, S. Rahmstorf, R. Ramesh, D. Raynaud, D. Rind, O. Solomina, R. Villalba, and D. Zhang, "Palaeoclimate," in S. Solomon, D. Qin, M. Manning, Z. Chen, M. Marquis, K. B. Averyt, M. Tignor, and H. L. Miller (eds.), *Climate Change 2007: The Physical Science Basis.* Contribution of Working Group I to the Fourth Assessment Report of the Intergovernmental Panel on Climate Change (Cambridge: Cambridge University Press, 2007), 433–498.

Kirtman, B., S. B. Power, J. A. Adedoyin, G. J. Boer, R. Bojariu, I. Camilloni, F. J. Doblas-Reyes, A. M. Fiore, M. Kimoto, G. A. Meehl, M. Prather, A. Sarr, C. Schär, R. Sutton, G. J. van Oldenborgh, G. Vecchi, and H. J. Wang, "Near-term Climate Change: Projections and Predictability," in T. F. Stocker, D. Qin, G.-K. Plattner, M. Tignor, S. K. Allen, J. Boschung, A. Nauels, Y. Xia, V. Bex, and P. M. Midgley (eds.), *Climate Change 2013: The Physical Science Basis.* Contribution of Working Group I to the Fifth Assessment Report of the Intergovernmental Panel on Climate Change (Cambridge and New York: Cambridge University Press, 2013), 953–1028.

Lemke, P., J. Ren, R. B. Alley, I. Allison, J. Carrasco, G. Flato, Y. Fujii, G. Kaser, P. Mote, R. H. Thomas, and T. Zhang, "Observations: Changes in Snow, Ice and Frozen Ground," in S. Solomon, D. Qin, M. Manning, Z. Chen, M. Marquis, K. B. Averyt, M. Tignor, and H. L. Miller (eds.), *Climate Change 2007: The Physical Science Basis.* Contribution of Working Group I to the Fourth Assessment Report of the Intergovernmental Panel on Climate Change (Cambridge: Cambridge University Press, 2007), 377–384.

Lisiecki, L. E., and M. E. Raymo, "A Pliocene-Pleistocene Stack of 57 Globally Distributed benthic delta O-18 Records," *Paleoceanography* 20 (2005), PA1003, doi: 10.1029/2004PA001071.

Marcott, S. A., J. D. Shakun, P. U. Clark, and A. C. Mix, "A Reconstruction of Regional and Global Temperature for the Past 11,300 Years," *Science* 339 (2013), 1198–1201, doi: 10.1126/science.1228026.

Meehl, G. A., C. Tebaldi, G. Walton, D. Easterling, and L. McDaniel, "Relative Increase of Record High Maximum Temperatures Compared to Record Low Minimum Temperatures in the U.S.," *Geophysical Research Letters* 36 (2009) L23701, doi: 10.1029/2009GL040736.

Myhre, G., D. Shindell, F.-M. Bréon, W. Collins, J. Fuglestvedt, J. Huang, D. Koch, J.-F. Lamarque, D. Lee, B. Mendoza, T. Nakajima, A. Robock, G. Stephens, T. Takemura, and H. Zhang, "Anthropogenic and Natural Radiative Forcing," in T. F. Stocker, D. Qin, G.-K. Plattner, M. Tignor, S. K. Allen, J. Boschung, A. Nauels, Y. Xia, V. Bex, and P. M. Midgley (eds.), *Climate Change 2013: The Physical Science Basis*. Contribution of Working Group I to the Fifth Assessment Report of the Intergovernmental Panel on Climate Change (Cambridge and New York: Cambridge University Press, 2013), 659–740.

National Research Council, *Surface Temperature Reconstructions for the Last 2,000 Years*. Board on Atmospheric Sciences and Climate (Washington, DC: National Academies Press, 2006).

Nerem, R. S., D. P. Chambers, C. Choe, and G. T. Mitchum, "Estimating Mean Sea Level Change from the TOPEX and Jason Altimeter Missions," *Marine Geodesy*, 33 (sup. 1) (2010), 435–446.

Nordhaus, W., *A Question of Balance: Weighing the Options on Global Warming Policies* (New Haven, CT: Yale University Press, 2008).

Petit, J. R., J. Jouzel, D. Raynaud, N. I. Barkov, J. M. Barnola, I. Basile, M. Bender, J. Chappellaz, M. Davis, G. Delaygue, M. Delmotte, V. M. Kotlyakov, M. Legrand, V. Y. Lipenkov, C. Lorius, L. Pepin, C. Ritz, E. Saltzman, and M. Stievenard, "Climate and Atmospheric History of the Past 420,000 Years from the Vostok Ice Core, Antarctica," *Nature* 399 (1999), 429–436.

Royer, D. L., "CO2-Forced Climate Thresholds during the Phanerozoic," *Geochimica et Cosmochimica Acta* 70 (2006), 5665–5675.

Solomon, S., D. Qin, M. Manning, R. B. Alley, T. Berntsen, N. L. Bindoff, Z. Chen, A. Chidthaisong, J. M. Gregory, G. C. Hegerl, M. Heimann, B. Hewitson, B. J. Hoskins, F. Joos, J. Jouzel, V. Kattsov, U. Lohmann, T. Matsuno, M. Molina, N. Nicholls, J. Overpeck, G. Raga, V. Ramaswamy, J. Ren, M. Rusticucci, R. Somerville, T. F. Stocker, P. Whetton, R. A. Wood, and D. Wratt, "Technical Summary," in S. Solomon, D. Qin, M. Manning, Z. Chen, M. Marquis, K. B. Averyt, M. Tignor, and H. L. Miller (eds.), *Climate Change 2007: The Physical Science Basis*. Contribution of Working Group I to the Fourth Assessment Report of the Intergovernmental Panel on Climate Change (Cambridge: Cambridge University Press, 2007), 19–92.

Solomon, S., G.-K. Plattner, R. Knutti, and P. Friedlingstein, "Irreversible Climate Change due to Carbon Dioxide Emissions," *Proceedings of the National Academy of Sciences* 106 (2009), 1704–1709.

Stanton, E. A., and F. Ackerman, *Florida and Climate Change: The Costs of Inaction* (Boston: Tufts University, 2007; available online at http://ase.tufts.edu/gdae/Pubs/rp/FloridaClimate.html).

van Vuuren, D. P., et al., "The Representative Concentration Pathways: An Overview," *Climatic Change* 109 (2011), 5–31, doi: 10.1007/s10584-011-0148-z.

Vaughan, D. G., J. C. Comiso, I. Allison, J. Carrasco, G. Kaser, R. Kwok, P. Mote, T. Murray, F. Paul, J. Ren, E. Rignot, O. Solomina, K. Steffen, and T. Zhang, "Observations: Cryosphere," in T. F. Stocker, D. Qin, G.-K. Plattner, M. Tignor, S. K. Allen, J. Boschung, A. Nauels, Y. Xia, V. Bex, and P. M. Midgley (eds.) *Climate Change 2013: The Physical Science Basis.* Contribution of Working Group I to the Fifth Assessment Report of the Intergovernmental Panel on Climate Change (Cambridge and New York: Cambridge University Press, 2013).

Weart, S. R. The Discovery of Global Warming, 2nd ed. (Cambridge, MA: Harvard University Press, 2008). Accessible online at www.aip.org/history/climate/index.htm

Zachos, J., M. Pagani, L. Sloan, E. Thomas, and K. Billups, "Trends, Rhythms, and Aberrations in Global Climate 65 Ma to Present," *Science* 292 (2001), 686–693.

# Index